Practitioner's Handbook of Risk Management for Water & Wastewater Systems

Practitioner's Handbook of Risk Management for Water & Wastewater Systems

Gordon Graham and Paul Fuller

CRC Press is an imprint of the
Taylor & Francis Group, an **informa** business

First edition published 2022
by CRC Press
6000 Broken Sound Parkway NW, Suite 300, Boca Raton, FL 33487-2742

and by CRC Press
2 Park Square, Milton Park, Abingdon, Oxon, OX14 4RN

Library of Congress Cataloging-in-Publication Data

Names: Graham, Gordon 1951- author. | Fuller, Paul (Insurance Professional), author.
Title: Practitioner's handbook of risk management for water & wastewater systems / Gordon Graham and Paul Fuller.
Other titles: Practitioner's handbook of risk management for water and wastewater systems
Description: First edition. | Boca Raton : CRC Press, [2022] | Includes index.
Identifiers: LCCN 2021027700 (print) | LCCN 2021027701 (ebook) | ISBN 9781032133898 (hbk) | ISBN 9781032134130 (pbk) | ISBN 9781003229087 (ebk)
Subjects: LCSH: Water treatment plants--Risk management. | Water treatment plants--Employees--Protection. | Water--Purification--Safety measures | Sewage--Purification--Safety measures
Classification: LCC TD358 .G73 2022 (print) | LCC TD358 (ebook) | DDC 628.1/62--dc23
LC record available at https://lccn.loc.gov/2021027700
LC ebook record available at https://lccn.loc.gov/2021027701

ISBN: 978-1-032-13389-8 (hbk)
ISBN: 978-1-032-13413-0 (pbk)
ISBN: 978-1-003-22908-7 (ebk)

DOI: 10.1201/9781003229087

Typeset in Times
by Deanta Global Publishing Services, Chennai, India

Disclaimer

The information contained in this book is meant to provide the reader with a general understanding of certain aspects of risk management. The information isn't meant to be construed as legal, risk management, or insurance advice and isn't meant to be a substitute for legal, risk management, or insurance advice. AAWD&M, Gordon Graham, and Paul Fuller expressly disclaim any and all liability with respect to any actions taken or not taken based upon the information contained in the book or with respect to any errors or omissions contained in such information.

Contents

Foreword

I wish that *real* risk management had been written forty-five years ago when I was beginning my career in the water industry. Authors Gordon Graham and Paul Fuller distill years of risk management experience into a thoughtful book that delivers real insights into the practice with enough personality and humor to keep me turning pages. Full disclosure: This book made me realize how little I knew about *real* risk management and how valuable it is to any organization.

I'll admit it: I understood the importance of workplace safety but I didn't fully grasp what *real* risk management was until I read this book. That's hard to own up to after a career that started as a serviceman for a water utility then progressing up the management ladder first as a crew leader, then foreman, superintendent, director of operations, and now president of California Domestic Water Company and an elected board member to the Hi-Desert Water District in Yucca Valley, California. I see now how fortunate I was throughout my career to have avoided tragedy or casualty at work. And I also now see how *real* risk management could be incorporated into safety trainings for the benefit of all.

I've had the pleasure of hearing Gordon speak in person. Now, some people might think insurance and risk management experts are dull. They picture someone in a grey suit delivering remarks in a monotone. Gordon defies those stereotypes in person and on the pages of this book, making hours fly by like minutes. Indeed, Gordon is that rare individual with a true gift and ability to turn any conversation into a learning experience. I think he was born with an intuitive sense for risk management, but he has the credentials too—beginning his career as a motorcycle officer with the California Highway Patrol, earning a master's degree from the prestigious Institute of Safety & Systems Management at USC and a law degree from Western State University of Law. He instinctively understands that *things that go wrong are predictable, and predictable is preventable.* Readers will find more of his terrific insights in the pages that follow.

Paul does a great job at capturing the essences of Gordon's insights and developing them further with practical tools and steps that readers will be able to put to work the next day. Smart and driven, Paul aims to help others improve their practices and leadership to put *real* risk management at the center. It's hard to believe this is his first book because he deftly captures Gordon's passion, personality, and wit and weaves together stories from his life experiences with the ten rules of *real* risk management. The result? A book that provides water and wastewater managers with the tools to prevent problems from becoming tragedies. Paul's writing style provides risk-managing practitioners the ability to instill into an organization the pursuit to *do the right thing, the right way every time.*

If you're like me, this book will enlighten you and inspire new thinking and dialog at your organization. I hope you enjoy the journey as much as I did.

Jim Byerrum **Chairperson, CalMutuals JPRIM**

Authors

Gordon Graham is a forty-plus-year veteran of law enforcement. His education as a risk manager and experience as a practicing attorney, coupled with an extensive background in public safety, has distinguished Mr. Graham as a nationally recognized and sought-after speaker for public- and private-sector professions with multiple areas of expertise. Gordon is a product of the greatest generation. Raised in San Francisco in the 1950s, he not only learned that love of God, country, and family were critical, but he was also taught the immense value of continuous learning and hard work as well as the importance of doing the right thing, the right way every time. These beliefs have been a constant in his life. After his first twelve years of formal education in the Catholic school system, Gordon began his undergraduate work at San Francisco State College during the tenure of S.I. Hayakawa who further reinforced those core values. He graduated in 1973 with a Bachelor of Arts in Business.

In 1973, Gordon was selected as a candidate for the California Highway Patrol (CHP). Thereafter, he proudly served as a motorcycle officer for most of his first ten years in the Los Angeles area. In addition to his patrol work, Gordon helped design the first driving under the influence (DUI) task force, assisted in the development of the drug recognition expert program (DRE), and was an instructor in the initial "Mod I and II" hazardous materials program. During this period, he wrote the first of his many technical papers, *PCP–An Officer's Survival Guid*e. Simultaneously, he was furthering his formal education by spending two years at Long Beach State College under the tutelage of Richard Kaywood where he received a lifetime teaching credential.

Following this degree, Gordon attended the Institute of Safety & Systems Management (ISSM) at the University of Southern California. He'll quickly tell you that ISSM was the best education he ever received from the brightest people in the field of risk management. His professors included Chaytor Mason, Ted Ferry, Bill Petak, and Harry Hurt. Gordon's relationship with Professor Hurt led to his collecting data for *The Hurt Report*. Published in 1980, this report on motorcycle fatalities was and is recognized as the single greatest treatise on motorcycle safety. After completing his Master of Science, Gordon's off hours were spent at Western State University School of Law, where he graduated in 1982 with his Juris Doctorate. He passed the California bar exam the same year and opened his law offices in Hollywood, where he focused on family law, immigration, and personal injury at work.

In his law enforcement life, Gordon was promoted to sergeant in 1982 and supervised his former unit at the CHP. He and his fellow sergeants on "B" shift stressed the values and beliefs of the CHP and built one of the most productive motorcycle officer teams in the history of the organization. As a sergeant, Gordon saw deficiencies in how officers were trained and revolutionized law enforcement training with his SROVT program (solid, realistic, ongoing, and verifiable training). This daily training concept resulted in his receiving the governor's award for excellence in law enforcement training. Gordon's personal life took a change for the better when he married his lovely bride, Reneé, in 1983. He was later promoted to a CHP headquarters position in 1992 where he continued his work in law enforcement training. During this period, Gordon sharpened his training curriculum by integrating components of human behavior, systems, risk management, and law. The outcome was a philosophy called *real* risk management and is applicable to any profession.

Gordon has delivered training programs to a variety of professions, including commercial real estate, public health, manufacturing, distribution, construction, aerospace, nuclear power, and auto racing. In 2002, he expanded his public safety focus by collaborating with Chief Billy Goldfeder to develop www.firefighterclosecalls.com. This reporting repository is recognized as the national source for firefighter safety data. It subsequently led to his assisting the International Association of Fire Chiefs (IAFC) with a similar effort, www.firefighternearmiss.com. The IAFC awarded Gordon with their presidential award for excellence for his lifelong work improving firefighter safety and performance.

That same year, Gordon, along with Bruce Praet, founded Lexipol, a company designed to standardize policy, procedure, and training for public safety organizations. Over two-thirds of the states in our great country use Lexipol's knowledge management system, including most California law enforcement agencies. This effort has improved the protection of our great public and the overall safety of police personnel. It has also influenced a reduction in adverse claims, settlements, and legal verdicts. Gordon held his CHP headquarters position until his service retirement in 2006. He has been awarded the lifetime achievement award from California Peace Officer Standards & Training (POST), the lifetime dedication award from the International Public Safety Leadership and Ethics Institute (IPSLE), and the James Oberstar sentinel of safety award for his lifetime work improving aviation safety internationally.

Today, Gordon divides his time among study, research, writing, speaking, and consulting in the field of risk management. His innovative programs, based on the values and beliefs he learned as a child, along with a passion for continuous improvement, are the standard for high-reliability organizations seeking to prevent tragedies through self-reinforcing systems, culture, and discipline. He and Mrs. G are the proud parents of two children who've taken these values and beliefs along with them in their personal and professional lives. A resident of Southern California, Gordon enjoys spending his leisure time with Mrs. G, cruising the coast in their vintage boat or car and exploring new adventures and new restaurants.

Additional details on Mr. Graham can be found on the following links:

www.gordongraham.com

www.lexipol.com

www.aawdm.org

Paul Fuller is an insurance professional specializing in public water and wastewater systems. He's Director of the American Association of Water Distribution & Management (AAWD&M), CEO of Allied Public Risk, LLC (a national managing general underwriter, exclusively focused on public entities and public water and wastewater systems), and Insurance Administrator for the California Association of Mutual Water Companies Joint Powers Risk and Insurance Management Authority (CalMutuals JPRIMA). Throughout his twenty-five-year career, Paul has underwritten thousands of public water and wastewater systems. He has also overseen tens of thousands of their claims, with emphasis on inverse condemnation, water quality, coverage disputes, and property valuations.

Mr. Fuller earned his Bachelor of Business Administration from the University of San Diego and Master of Science in Risk Management and Insurance from Georgia State University. His past industry experience includes: President, Alteris, Inc.; Wholesale and Specialty Programs Operations President, Glatfelter Insurance Group; and President, S.N. Potter Insurance Agency, Inc. He holds the Chartered Property Casualty Underwriter (CPCU) designation, Certified Public Risk Officer-Water/Wastewater (CPRO-W^2) designation, California Water Operator license (D2/T1), California Qualified Claims Manager license, and is nationally licensed as a Property and Casualty Agent and Surplus Lines Broker. Paul is a frequent speaker on risk management and litigation issues. His articles on inverse condemnation have been published in the *Public Law Journal* and *Defense Counsel Journal*. He's also the author of *Part IV. Insurance Essentials* for the Certified Public Risk Officer-Water/Wastewater (CPRO-W^2) designation.

Additional details on Mr. Fuller can be found on the following links:

www.alliedpublicrisk.com

www.aawdm.org

https://www.linkedin.com/in/paul-fuller-cpcu-b56a8612

Introduction

TEN REASONS TO READ THIS BOOK

****

I don't need to read this book because we have a great safety program in our organization.

As you move through the various chapters, you'll quickly learn that having a great safety program is important, but *real* risk management is much more than safety stuff. Everything you and your people do involves a level of risk.

****

I'm sure the authors are nice, well-meaning people, but bad things are just going to happen, and there really are no practical control measures to prevent this reality.

You're right. The authors are nice, well-meaning people. Having clarified that point, bad things don't have to happen, and, in reading this book, you'll learn proven strategies and tactics to identify potential problems and address them proactively.

****

In our state, the unions are so powerful that nothing can be done to make things better in our organization.

What's the primary mission of an employee union? It's protection of membership, and what better way to protect them from physical harm, discipline, or lawsuits than *real* risk management? It all gets down to managing risk.

****

With all the financial issues we're facing right now, we don't have any room in our budget to make our risk management program better.

As you proceed through the chapters, you'll quickly learn that most of these proven strategies can be implemented at either no or little cost. On the other hand, your efforts to bring these strategies into your workplace have the potential to save you millions of dollars.

****

We really don't need a formal decision making process because most of what our people do, they're doing right.

You're right! Most of what your people do, they're doing right *because* they've done it before many times, and experience is a great teacher. What happens when good people get involved in a low-frequency event? Have you provided them with a proven system on how to think things through, or do you just hope they make the right call?

****

There's no need for me to cover harassment with my personnel because we don't have a problem, and everyone gets along well.

Both the authors of this book smile internally when people think this way! Yes, everyone gets along famously until they don't get along. Or, until someone finally becomes exacerbated with the given inappropriate behavior. Or, until someone is disciplined, and, all of a sudden, the welcome behavior turns into an unwelcome behavior.

****

I'm certain we don't have any problems with our current performance evaluation system because we've never had anyone complain about their ratings.

You're right! People don't complain when they're being overrated every year. Too many supervisors lack the guts to honestly evaluate people, and the inflated documents can come back to haunt you in ways you've never even imagined.

****

We perform a criminal history check on each applicant so we have the topic on background investigations already covered.

Comprehensive background investigations entail much more than a criminal history check. Always remember that nineteen of the nineteen hijackers on September 11, 2001, were devoid of a criminal history in the United States.

****

Our supervisors do a great job for us, and they get along well with their teams.

Here's a quick test you can give the next supervisor you see. "What's the primary mission of a supervisor in this organization?" If the answer is anything other than enforcement of organizational policy, you have a problem lying in wait. If people want to get along, then they should not promote.

****

We face thousands and thousands of risks in our operations. Where do we get started?

Early in this book, you'll be exposed to the ten families of risk. Your mission is to identify the three greatest risks in each of these families and to confirm you have viable control measures in place to properly address those risks.

1 Real Risk Management

Doing the Right Thing, the Right Way Every Time

2019© Photo Courtesy of Placer County Water Agency, Placer County California 95604.

STRATEGIC HINT FOR YOUR CONSIDERATION

Bias acts as a blind spot when evaluating problems lying in wait, limits your insight when recognizing and prioritizing risks, narrows your perspective, and leads to incomplete and inaccurate decision making.

DOI: 10.1201/9781003229087-1

SUMMARY

Real risk management is a belief structure of high-reliability organizations to do the right thing, the right way every time, through resilient systems and a culture of ongoing improvement, sound decision making, and self-regulating employee discipline.

The philosophy's underpinning is the prevention of tragedies by identifying and addressing problems lying in wait, as opposed to proximate cause, and converting these root causational factors into pillars of success. This conversion maximizes employee fitness, customer loyalty, and organizational mission.

Chapter emphasis will focus on pre-incident prevention and the importance of building resilient systems through enhanced perspective, data, and workplace diversity, as well as the necessity to monitor bias so that predictable gray rhinos are clearly distinguished from unpredictable black swans.

Gordon Graham here, and welcome to Chapter 1 of our journey toward *real* risk management. Throughout this trek, we'll be imparting practical ideas, strategies, tactics, and concepts that you can take back and implement in your organization. The book's primary purpose is to provide an actionable manual to better protect yourself, your organization, your community, your profession, and the big one for me: our great country. I'll periodically refresh salient topics as memory markers and include directional signs through strategic hints to ensure we reach our final destination of doing the right thing, the right way every time, which is the marque of a high-reliability organization.

These same reminders will apply to certain human conditions and traits that often transform problems lying in wait into tragedies. Such characteristics include bias and mediocrity, as well as arrogance, ignorance, and complacency. It also includes decision making that lacks data and diversity as well as behavioral tendencies that fixate attention on superfluous risks. Real risk managers understand an informed and varied perspective is necessary to properly scan for perceptible and imperceptible risks, and these practical actions lead to viable control measures (systems) that prevent tragedies.

Let's begin our journey with a four letter word that surfaces when I discuss risk management: *bias*. I give a couple hundred presentations a year, and, occasionally, I say things that make people angry. The common reaction is hate mail, and this reaction is exponentially greater when I publish articles or musings. I wrote a piece about a decade ago titled *Everyone Has a Bias*. Of all the pieces I've written, this one generated more hate mail than anything else. "Dear Mr. Graham: How dare you accuse me of having a bias! You don't even know me, and I can assure you that I don't have a bias bone in my body." I sent an email back. Not only are you biased, you're unwise if you don't recognize your own bias.

STRATEGIC HINT FOR YOUR CONSIDERATION

The best way to improve as a person and risk manager is to expand your horizons (perspective), and that's properly achieved through active and ongoing engagement with insightful books and smart people.

There's nothing wrong with bias. If you prefer Ford's to Chevy's, that's a bias. If you salt your food before you eat it, that's a bias. If you prefer one football team to another, that's a bias. There's nothing wrong with bias until it negatively affects people or narrows perspective. Bias, from the lens of risk management, adversely influences our decision making and ability to properly evaluate risks. Let me explain. A common bias I face when speaking to audiences outside of public safety is the following: "What's this guy talking about? Number one: He's not experienced in our industry. Number two: What does he know about my job? And number three: What were they thinking when they asked a non-industry guy to talk about safety stuff?" That level of bias is myopic and contributes to the mishandling of risk. It also acts as a blind spot when evaluating problems lying in wait and constricts your insight when recognizing and prioritizing organizational and operational risks. All those factors narrow your perspective and lead to incomplete or inaccurate decision making. That brings me to a central point in my philosophy: The best way to improve as a person and risk manager is to expand your horizons (perspective), and that's best achieved through active and ongoing engagement with insightful books and smart people.

Let me drilldown further. What do I specifically mean when I say perspective? I'm referring to a proper and expansive comprehension of this broad, multifaceted field known as risk management. It's disconcerting but most professionals have a lack of understanding of this field. Ninety-nine percent of people throughout this great country view risk management as safety stuff. I've ten chapters and an illustrative conclusion to convince you it's more than safety stuff. Everything you do in water/wastewater operations, regardless of your job, involves a level of risk. For the executives reading this book, you're hiring people. There's a level of risk. I said that recently and somebody challenged me: "That really doesn't apply to me. I only have three employees." And my response to that? Three plaintiffs and three defendants. You're firing somebody. There's a level of risk. "Well, I can terminate an employee for any reason I want." You are mostly right. In some states, you can terminate an employee for no reason or any reason. However, there isn't a state where you can terminate an employee for the wrong reason. There's a level of risk. You're completing a performance evaluation. There's a level of risk. You're mixing chemicals. There's a level of risk. You're shoring up a trench. There's a level of risk. Every activity performed by your organization involves a level of risk.

STRATEGIC HINT FOR YOUR CONSIDERATION

High-reliability organizations intrinsically embody real risk management as a dynamic, inculcated way of thinking that's predicated on their mission/purpose and serves as a governing luminance for their employees.

The purpose of this exercise is to show you the breadth and depth of what I call *real* risk management. Not the unctuousness I observe in so many organizations but a commitment to *real* risk management. I'm dedicated to imparting practical tools that you can apply in your organization and personal life. I also want to correct any misperception you have about risk management, and why that incomplete perspective impacts the activities you perform. *Real* risk management goes beyond the safety stuff. High-reliability organizations intrinsically embody *real* risk management as a dynamic, inculcated way of thinking. It isn't a siloed department but a manifestation of an enterprise's mission and purpose, as well as a governing luminance for employees. *Real* risk management is an indefatigable pursuit to do the right thing, the right way every time. It's the marrow of high-reliability organizations, with systems and culture serving as genetic material and discipline functioning as plasma. *Real* risk management is the reason these marque organizations are underrepresented in tragedies (i.e., injuries or death to personnel, civil liability, organizational embarrassment, internal investigations, and criminal filings) and exceed their peer groups in employee fitness and customer loyalty. When you do the right thing, the right way every time, you maximize service, enhance safety, and minimize liability. That's the positive and far-reaching impact of *real* risk management. And, it's easy to deploy if you commit yourself to improving your bias and expanding your perspective.

After just a few pages, you've likely surmised I'm a risk management enthusiast. It's been my life's mission and life's work for forty-plus years. I'm frequently asked how I ended up in this line of work and how I became so passionate. I'd be disingenuous if I said it was a master plan that began in my twenties. It all started on August 4, 1977. I was a motorcycle cop for the California Highway Patrol (CHP) working in Central Los Angeles, and we had an event that date called Barz versus Burkholder. Kurt Barz was a sergeant in the Los Angeles Police Department (LAPD). I can't say we were friends, but he seemed like a nice enough guy when we occasionally crossed paths at the divisional detention facility. Barz was driving around in the early morning hours on August 4 and came into contact with Ronald Burkholder, a short stature man who was naked and karate chopping a phone booth. Barz noticed the unusual behavior and exited his Rambler. According to witnesses, Burkholder immediately attacked and overpowered Barz. He then took Barz's baton and started hitting him with it. Shortly thereafter, Barz shot Burkholder six times and killed him.

STRATEGIC HINT FOR YOUR CONSIDERATION

Real risk management is the reason high-reliability organizations are under-represented in tragedies and exceed their peer groups in employee fitness and customer loyalty.

The public was outraged. It was the twenty-seventh fatal police shooting by LAPD so far that year. The friends of Burkholder were outraged: "How could you shoot him? He was a gentle soul, a PhD, and never raised his hands in anger to anyone." The American Civil Liberties Union (ACLU) was outraged. Everybody was outraged. Three days later, the coroner, Thomas Noguchi, came back and said: "New development: PCP." Now everybody reading this chapter has heard of PCP, but that wasn't the case in 1977. I'd never heard of PCP. Apparently, I wasn't alone because every night on the local news there were experts, MDs and PhDs, talking about PCP being a powerful animal tranquilizer. PCP makes you impervious to pain, gives you superhuman strength, turns you crazy, and makes you so hot you get naked. I'm watching all these experts on television and saying to myself: Wait a minute. Barz is a big guy, and Burkholder was a small guy, but Barz was overpowered and had to shoot Burkholder. What happens to Gordon Graham who isn't all that big or all that tough? What happens if I run into a tough guy on PCP? How much trouble am I going to get in? I'd better learn about PCP.

So I wrote down all the names of the experts on television. The next day when I arrived at work, I looked them up in the Yellow Pages. For my younger readers, the Yellow Pages was a big book with all sorts of phone numbers. I also wrote down their addresses, and the next day I drove out to their offices in uniform on my CHP motorcycle. I asked for a few minutes of her or his time because I had questions about PCP. They were kind and agreed to talk to me. They explained about PCP, and I wrote down everything they said. And for some reason, I took my accumulated notes and assembled them into a booklet that was eight pages long and the size of a ticket book. I called it, *PCP: An Officer's Survival Guide*. The booklet had three sections. Section one: How to identify these idiots on PCP? Section two: Here's what you don't do when you run into one of these fools on PCP. And section three: Here's what you need to do when you run into one of these PCP guys. I made forty copies of this booklet and distributed it to all the motorcycle cops on afternoon "B" shift in Central Los Angeles where I was working. You need to learn about PCP. It can be dangerous dealing with someone on PCP. You need to read this booklet. I handed it out and thought that was it.

STRATEGIC HINT FOR YOUR CONSIDERATION

The positive and far-reaching impact of real risk management includes maximizing service, enhancing safety, and minimizing liability; it's attainable if you address your bias and expand your perspective.

Thirty days later, my captain called me into his office: "Hey, Graham." Yes, captain. "Did you write this booklet?" I did. "I want a copy of it for every officer in the office, and I want you to go to all three shift briefings and explain PCP. We're not going to have a PCP tragedy in Central Los Angeles." So, here I am. A young motorcycle cop going to day "A" shift and talking to all the sixty-year-old motorcycle cops about PCP. I then go to graveyard "C" shift and talk to all the new hires from the Academy about PCP. And what do you learn when standing in front of people? You pick up platform skills to get people interested. People want to learn, and I shared my understanding of PCP to all three shifts. One month later: "Hey, Graham?" Yes, captain. "The division commander is concerned about PCP. I informed him that we have the subject matter expert in our office." Who's that? "That would be you. He wants you to go to every office in Southern Division and explain PCP." So now I compiled a slide tray (for my younger readers, that's a projector with photographs) and included pictures of people on PCP, some of whom were dead, from Los Angeles Police Department and Los Angeles Sheriff's Office. With my slide tray, I was traveling all over Los Angeles County talking about PCP and developing invaluable platform skills. I subsequently did all the offices in Southern California.

One month later: "Hey, Graham." Yes, captain. "The Academy's doing a videotape on PCP, and they want you to narrate it." I flew up to headquarters and did a videotape on PCP, which was distributed statewide to all public safety organizations. And that's when the phone started ringing: "Are you Gordon Graham?" Yes. "This is the chief of Riverside Police Department." Hey, chief. "Are you the guy on PCP?" Well, I'm not on it, but I talk about it. "How much do you charge?" Charge? I hadn't thought about that. So now you know how it all started in 1977. But there's more to the story. I had just finished graduate school in August 1977. In September, I enrolled in law school and graduated in 1982. I then opened up my law practice in Hollywood. Since 1982, I've been a lawyer, and as a lawyer, I've been handling tragedy. For almost forty years, people have been coming into my law office sharing their tragedies and requesting legal representation. "My son died in a motorcycle collision." "My house was destroyed by a pipeline explosion." "I'm being indicted for excessive force." "I'm being fired for harassment." "My husband died in a plane crash." For almost forty years as a lawyer, I've been immersed and fixated on tragedy. It has become my passion and purpose in life.

STRATEGIC HINT FOR YOUR CONSIDERATION

Most of your tragedies (i.e., injuries or death to personnel, civil liability, organizational embarrassment, internal investigations, and criminal filings) are caused by errors from your good employees and can be proactively addressed.

My story isn't quite finished. Remember, I mentioned attending graduate school before law school. And where did I attend graduate school? The Institute of Safety

& Systems Management (ISSM) at University of Southern California (USC). Where did ISSM come from? Our US military. Did you know our US military once had a terrible safety record? Training injuries during World War II were devastating. For those of you who haven't read the book *Unbroken: The Story of Louie Zamperini*, please pick up the book. I read *Unbroken*, and one piece of data concerned me. The US Army Air Corps from 1942 to 1945 was training pilots, and these pilots were suffering nineteen fatalities per day in training exercises. At the end of World War II, the military acknowledged their safety problem. In 1952, the military, primarily the Air Force at that time, decided to do something about it and contracted with USC to educate military leaders on the principles of risk management. Twenty or so years later, they opened ISSM to non-military personnel. I enrolled in one of the first classes for non-military personnel, and it was there that I became consumed in the study of tragedies and how to prevent them. ISSM taught me the skills that I'm imparting to you and is the foundation of my philosophy on *real* risk management. Figures 1.1 and 1.2 encapsulate the essence of this book.

Let's dissect Figure 1.1. As a lawyer, which is the right side of the graphic, I handle tragedies. As a risk manager, which is the left side of the graphic, I study tragedies and look for their root causational factors. My risk management journey has introduced me to scores of occupations, professions, and tragedies. I don't just talk to public safety professionals and water/wastewater professionals. I assist the timber industry, second most risky job in the United States. I assist the construction industry, third most risky job. I assist the trucking industry, fourth most risky job. I also assist owners of refineries and operators of nuclear power plants. It doesn't make any difference which high-risk profession you select. If you can identify what caused the tragedy (i.e., the root causational factors or problems lying in wait), then you can build actionable control measures (systems) to prevent future similar tragedies.

Organizational Risk Management

High-reliability organizations build resilient systems that are properly designed, kept up-to-date, and fully implemented. These systems proactively prevent tragedies by addressing problems lying in wait, which are the root causational factors. Successful systems require recognizing and prioritizing risks through ongoing and data-centric risk assessments and risk/frequency matrixes as well as embedding an organizational culture of employee discipline, continuous improvement, and system conformance.

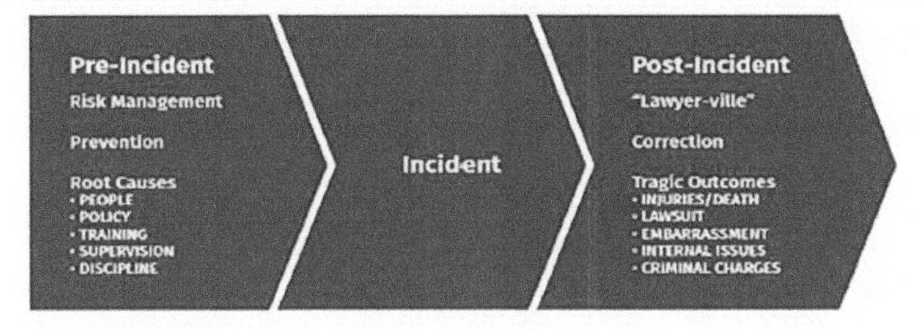

FIGURE 1.1 Organizational risk management. Source: Institute of Safety & Systems Management (ISSM).

***Real* Risk Management**

- There are specific techniques for addressing real problems lying in wait, and, when followed, they'll convert your root causational factors from problems to solutions via the five pillars of success.
- *Real* risk management is the process of looking into the future and identifying the real problems lying in wait so you can do something to prevent the occurrence (tragedy). This process involves recognizing risks your organization faces, prioritizing these risks in terms of potential frequency, severity, and available time to think, and mobilizing (acting) to do something about the recognized and prioritized risks via resilient systems that are properly designed, kept up to date, and fully implemented.
- *Real* risk management encompasses getting and keeping good people, deriving and maintaining good policies, confirming adequate training on policies, ensuring appropriate supervision of employees so policies are followed, and taking appropriate discipline when there's deviation from established policies.
- *Real* risk management has three fundamental precepts: (1) There's no new ways to get into trouble; (2) There's always a better way to stay out of trouble; and (3) Predictable risks are preventable risks, meaning identifiable risks are manageable risks.
- Most organizational tragedies are caused by employee error and, therefore, can be proactively addressed. These errors invariably involve activities that are high risk, low frequency, with either no time to think or time to think but the thinking is flawed. Employee errors also comprise activities that are high risk, high frequency where complacency, distraction, hubris, exhaustion, or risk homeostatis is the root cause of the error.
- The root causational factors for all internal errors comprise people, policy, training, supervision, and discipline. These problem factors can be converted to solution factors and form the five pillars of success.
- Organizations that follow the principles of *real* risk management are underrepresented in tragedy and overrepresented in employee fitness and customer loyalty.

FIGURE 1.2 *Real* risk management. Source: Institute of Safety & Systems Management (ISSM).

And there are five root causational factors that invariably cause your tragedies: people, policies, training, supervision, and discipline. When identified and addressed, these problem factors become solution factors and transform into pillars of success. Systems refer to control measures, policies, procedures, protocols, processes, rules, and checklists. Your systems are only effective if they're properly designed, kept up to date, and fully implemented.

Here's another reminder: There's always an elite class of organizations in dangerous industries that are underrepresented in tragedies. Do you recall why? These high-reliability organizations have committed themselves to *real* risk management and have built resilient systems around the following three precepts: (1) The errors you're going to make can be predicted from the errors already made; (2) there's always new ways to prevent errors through continuous system improvements; and (3) predictable is preventable, meaning identifiable risks are manageable risks. The foregoing precepts are enveloped in systems, culture, and discipline. Let's now transition to Figure 1.2, which summarizes my philosophy. Figures 1.1 and 1.2 embrace the lessons you'll learn from this book. It's all I know and reflects my training in graduate and law school. And let me digress, which I'll often do, to accentuate the point. I was flying back recently to Los Angeles and was sitting next to a nice guy. He wanted to talk about the stock market. "So what do you think about the stock market?" I don't have a clue. Prior to that, the guy sitting next to me wanted to talk about basketball.

"Do you have your final four picked?" I don't know anything about basketball. All I know are Figures 1.1 and 1.2.

STRATEGIC HINT FOR YOUR CONSIDERATION

In order to build resilient systems to prevent tragedies from occurring, you must identify the five root causational factors invariably involved in a given tragedy (people, policy, training, supervision, and discipline).

As a lawyer, I handle tragedies. As a risk manager, I study tragedies and then try to build viable control measures (systems) to prevent similar tragedies from reoccurring. That's my passion, and it requires risk management training to identify the root cause(s) of a given tragedy. If you've been to any of my classes or watched any of my videos, you know my annoyance. We don't actively teach risk management in vocational, trade, or academic schools. Underline this statement: Women and men are assigned complex, high-risk jobs like yours without any training on risk management.

How dare you put people in high-risk jobs without any training in risk management. Regularly, I say that, and, regularly, I'm challenged. "Oh no, Gordon, that's not true. We talk about risk management all the time." No we don't. We're uninformed about risk management. "Oh no, we're committed to risk management." Are you really? Then show me your organizational chart. You just told me you're committed to risk management. Where will I find risk management? If it's on there, I'm surprised, and I'm doubly surprised if it's anywhere near the top where it has influence. You know who gets top billing on every organizational chart? It's the lawyers whose bias is fixing problems after they occur or post-incident correction. They get top billing, and the people whose bias is preventing problems from occurring, or pre-incident prevention, aren't even on the chart. If they're on the chart, they're sharing a box somewhere in the middle without influence. I regularly see risk management with maintenance, or I see risk management with finance. And my favorite is risk management with human resources.

STRATEGIC HINT FOR YOUR CONSIDERATION

Systems (i.e., control measures, policies, procedures, protocols, processes, rules, and checklists) are only effective if they're properly designed, kept up to date, and fully implemented.

Another diversion, but these stories all accentuate the lack of comprehension for risk management and its resulting impact on bias and perspective. I'm talking to a woman on an airplane. She's a risk manager for a major city. A big one. She says: "I've been the city's risk manager for thirteen years, and I've never had a class on risk management." What's your job? "I'm the director of human resources, and one of my assigned duties is risk management." That proves it. We don't get the concept of risk management. Let's now transition to a past discussion with public safety officers. "Bad things are just going to happen. There's nothing we can do." I heard that phrase when I was talking to a group of cops. "Bad things are just going to happen. There's nothing we can do." Really? Let's study line-of-duty deaths in US law enforcement. My first full year was 1974, and we had two-hundred-eighty-three line-of-duty deaths. How many did we have in 2017? One-hundred-thirty-three. More than a fifty percent reduction, and don't tell me the level of violence has gone down against cops. The level of violence has significantly gone up. How'd we go from two-hundred-eighty-three to one-hundred-thirty-three? Two words: risk management. I had nothing to do with the outcome, but there were some smart people in public safety who identified the root causational factors behind line-of-duty deaths and built viable control measures (systems) to prevent these tragedies from self-repeating.

We know so little about risk management, and many of us don't know how to identify the cause. "Gordon that's not true. Identifying cause is easy." Really? For forty-plus years, I've been sharing stories of Titanic with my audiences. Why do I talk about Titanic? Wherever I go in the world, people know the story about Titanic: An ocean liner that sunk and killed fifteen-hundred people in 1912. That's a tragedy. What caused it? You walk up to one-hundred Americans, and ninety-nine will tell you what? "The iceberg. I saw the movie. The ship hit the iceberg. The ship went down. Therefore, the iceberg had to be the cause." Why do we do that? Why is it when we ask people what caused the tragedy, they default to the event in time that immediately preceded the tragedy? The ship hit the iceberg. The ship went down. Therefore, the iceberg had to be the cause. So what do we do? We build control measures (systems) to address icebergs, and ships continue to sink? "Well Gordon, are you suggesting the iceberg wasn't the cause?"

STRATEGIC HINT FOR YOUR CONSIDERATION

Anyone can identify the proximate cause of a tragedy, but you're dooming your organization to future tragedies if you're building control measures (systems) solely on proximate cause.

If you're fading out, please fade in for a moment. In any occupation, in any profession, in any tragedy, rarely is it a single event that generates the tragedy. Usually, it's a cascade of events over a period of time that goes uninterrupted. We have a triggering event in risk management, which we call proximate cause. Anyone can identify the proximate cause of a tragedy, but you're dooming your organization to future tragedies if you're building control measures (systems) solely on proximate cause.

Real risk managers, and that's where I want you headed after reading this book, don't stop with proximate cause. They go back in time and ask this question: Were there related causes, contributory causes, associated causes, conditions, cultures, and problems lying in wait (root causes) that everybody knew or should've known about but no action was taken? And what happens when we don't address these problems lying in wait? Sooner or later, all the holes in the Swiss cheese line up (we'll soon discuss this concept by James Reason), and we have the triggering event followed by the tragedy. At that point, the lawyers get involved, and they do the post-incident investigation on the lawsuit, death, injury, embarrassment, or other issue. They peel back the various layers and identify the perceptible problems lying in wait, sometimes for years, sometimes longer, where action was never taken but everyone knew or should've known the real issues behind the tragedy.

This concept of root causational factors, or problems lying in wait, is the essence of *real* risk management and must be completely understood. Let's use an example to crystallize our fluency between root cause and proximate cause. An emergency vehicle is en route to a low-level service call. The vehicle operator runs a red light and kills a child on a Vespa while simultaneously being ejected from the vehicle. This event is indeed a tragedy and comprises the death of a child, injury to your operator, organizational embarrassment, loss of reputation, internal investigation, civil liability, and possible criminal filing. What caused the tragedy? The default answer is the operator ran a red light. That event, which we call the proximate cause, triggered the tragedy but isn't the real cause of the tragedy. What's the real or root causational factors behind the tragedy?

STRATEGIC HINT FOR YOUR CONSIDERATION

Real risk managers always ask this question: Were there related causes, contributory causes, conditions, cultures, and problems lying in wait that everybody knew or should've known about but no action was taken?

The post-incident investigation laid out the following findings: *People*: The involved operator had an expired driver's license and was involved in four preventable collisions over the past three years. On his performance evaluations, he received accolades for good driving and "always setting the proper example." *Policy*: The manual the operator had access to in the report-writing room stated when responding to calls, always "proceed with due regard for your safety and safety of others." The manual didn't reference what to do when there's a red light. During the investigation, it was learned the chief recently modified the policy to address red lights, but it wasn't updated in the policy manual located in the station. *Training*: There hadn't been formal training on vehicle operations in the past three years. Supervisors "discussed it with personnel" during scheduled ride-alongs, but there wasn't any written documentation of that discussion. *Supervision*: The involved operator had a history of wanting to be first-in (the first person to get to the scene of the event). He also had a prior collision while rolling to a call. His supervisor said he discussed this bad driving with the employee several times. *Discipline*: Not one member of the department had ever been disciplined for an activity related to vehicle operations.

That, friends, is the difference between proximate cause and real or root causational factors behind the tragedy. It unequivocally underscores the necessity to proactively address problems lying in wait before they become tragedies. Now please take a moment to reflect on the financial, emotional, and reputational costs to the organization and all impacted parties. There are only losers in this example. Let's keep on this subject a bit longer. Why do we fail to address problems lying in wait and gravitate toward proximate cause? Mediocrity. Margaret Heffernan, who's on my recommended reading list, has a great book, *Willful Blindness*, exactly on this topic. She sums up the problem nicely. And I paraphrase: "Why don't people raise their hand and address the problems lying in wait?" Her answer was clear: "In too many organizations, if you raise your hand and say we have a problem, in the eyes of many people in the organization, you're the problem so you don't raise your hand." This phenomenon is acute in most organizations because if you raise your hand and take risk, you get paid X. If you don't do anything and ignore the risk, then how much do you get paid? The same amount of money. So why raise your hand and take the risk? As a result, in too many organizations, including too many water-related entities, mediocrity has replaced accountability. And when there's no accountability, the organization's integrity and purpose begin to fail. The derivatives of mediocrity are organizational tragedies, mission failure, and customer disaffection.

STRATEGIC HINT FOR YOUR CONSIDERATION

When there's no accountability, an organization's integrity and purpose begin to fail. The derivation is employee mediocrity, which exacerbates tragedies, mission failure, and customer disaffection.

STRATEGIC HINT FOR YOUR CONSIDERATION

Precept one: The errors you're going to make can be predicted from the errors already made. Although there are variations on a theme, it's essentially the same types of errors on repeat cycle.

What are you doing to eliminate mediocrity from your organization? You have a woman working for you, a fifteen-year employee who consistently shows up early and leaves late. A good person who can be counted on to do the right thing, the right way every time. All of a sudden, she wants to leave your organization and go someplace else. Should the bells of Saint Mary be going off in your head? Why is she leaving? Did you even ask? Or do you have a bias? "She's just disgruntled? They pay two hundred dollars a month more than we pay." Maybe she's aware of something nefarious going on in your organization? Perhaps theft, inappropriate behavior, racism, or sexism? If you ask the right questions, she might answer those questions. Or, do we just check the boxes to make sure all organizational kit is returned? Please take your exit interviews seriously. Are you doing your own audits? Do you have a robust formal and informal audit process to ensure what you say you're doing is in fact being done? If you're doing your own audits, then you have a built-in bias. Everything is perfect in your consciousness when in reality there are significant problems lying in wait. The solution correlates back to *real* risk management and its various tools, which will be imparted throughout this book.

To be fair, not all tragedies can be prevented. It's difficult to prevent intentional misconduct from external sources. However, that same misconduct from your own employees can be mitigated through proper recruitment, comprehensive background investigations (pre-hire and post-hire), effective probation protocol, meaningful performance evaluations, consistent discipline, and extensive audit processes. The vast majority of your tragedies (i.e., injuries and death to personnel, civil liability, organizational embarrassments, and internal investigations) are caused by mistakes, negligence, and errors from your good employees. These types of tragedies can be addressed proactively by reviewing your past tragedies and near misses and establishing appropriate control measures (systems) to prevent similar tragedies from reoccurring. This approach is based on the precept that past errors are predictive of future errors. What specific steps can you take to prevent tragedies in your high-risk, complex profession? It all gets down to *real* risk management. Let's review these specific tools in more detail, but first we must recall some fundamental definitions.

Time for a refresher. Risk is the possibility of meeting danger or suffering a harm or loss, or exposure to harm or loss. As a follow then, risk management is any activity that involves the evaluation of or comparison of risk and the development, selection, and implementation of control measures (systems) that change outcomes. Or more simply stated, risk management is the process of looking into the future (short or long term) and asking what can go wrong and then doing something about it to prevent it from going wrong. This process begins with a concept known as RPM (recognition, prioritization, and mobilization). First, you must recognize the risks facing your organization. Next, you must prioritize these risks in terms of potential frequency, severity, and available time to think. Finally, you must mobilize (act) to address the recognized and prioritized risks. The mobilization or action components of RPM are your systems. What's a system? The word gets thrown around a lot, but what does it mean? System refers to an organized or established procedure or an accumulation of processes. When you check the definition of process and procedure, it states a particular way of accomplishing something and a series of steps followed in a regular definite order. Systems that are properly designed, kept up to date, and fully implemented will never let your organization down.

Let's revisit employee errors, which we know are the cause of most organizational tragedies. James Reason (one of the great writers on risk management who originated the Swiss cheese model) describes three types of errors caused by employees: knowledge based; rule based; and skill based. All three are preventable. He goes on to examine how errors, lapses, omissions, and mistakes occur in any given workplace. Activities go wrong, not because we have bad people (there are specific risk management processes to eliminate bad people), but because our good people occasionally get involved in high-risk, low-frequency activities where there's no time to think (non-discretionary time), or where there's time to think (discretionary time), but the thinking is flawed. When these activities aren't done right, there's potential for tragedies. And that takes us to training. Take away frequency, and you've taken away experience. Take away experience, and you're only left with training. The key to doing the right thing, the right way every time is training. That's especially important for high-risk, low-frequency activities that involve no time to think (non-discretionary time). Figure 1.3 provides a primer to James Reason's Swiss cheese model. This model accentuates the value of resilient systems, continuous improvement, consistent decision making, and an organizational culture of self-regulating employee discipline.

Swiss Cheese Model of System Accidents

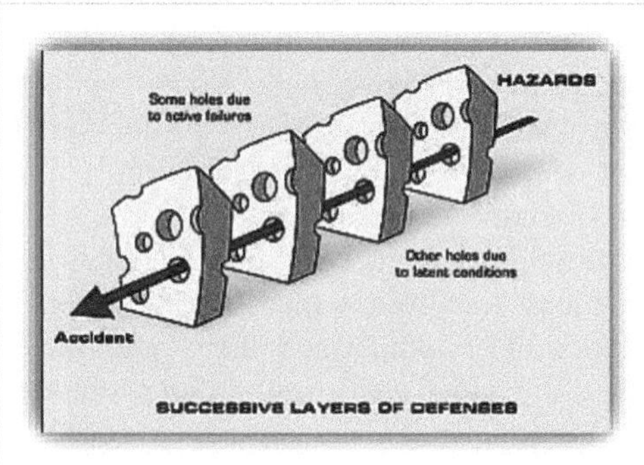

The Swiss cheese model is a useful method to explain the concept of cumulative effects. James Reason compares organizational systems to layers of Swiss cheese. Defenses within organizational systems occupy key layers. In an ideal world, each defensive layer would be intact. In reality, they're more like slices of Swiss cheese, having many holes, although, unlike in the cheese, these holes are continually opening, shutting, and shifting their location. The holes in the defenses arise for two reasons: active failures and latent conditions. Nearly all adverse events involve a combination of these two factors. Active failures are the unsafe acts committed by people who are in direct contact with the system, whereas latent failures include contributory factors that lie dormant until they contribute to the accident. Hazards when they travel through the Swiss cheese holes lead to accidents. Hazard Sequence: The first few slices are prevention or containment. The next slices are control or detection, which are leading indicators and comprise loss of primary containment if they travel through them. Next, we have mitigation and emergency, which are lagging indicators and create a major incident or consequence should it travel through them. The Swiss cheese model indicates that organizations must maintain a consistent mindset of intelligent wariness.

FIGURE 1.3 Swiss cheese model of system accidents. Source: James Reason Swiss cheese model. Source: BMJ, 2000 Mar 18:320(7237): 768–770; The James Reason Swiss cheese failure model in 300 Seconds by Whatsthepoint (May 30, 2018).

STRATEGIC HINT FOR YOUR CONSIDERATION

Precept two: There's always a better way to prevent tragedies through continuous system improvement and commitment of employees to embrace organizational mission, purpose, discipline, systems, and culture.

Providing training as close as possible to pre-incident is essential, and that requires development of systematic control measures (systems) to assure that your people are adequately trained. Otherwise, all the holes in the Swiss cheese will eventually align and tragedy will occur. The lawyers then take over and the problems lying in wait are identified and addressed after the fact. By then, it's too late. The same logic applies to errors in high-risk, high-frequency activities where complacency, distraction, hubris, fatigue, or risk homeostasis is the root cause of the tragedy. Preventing tragedies requires resilient systems that are kept up to date and fully implemented. The sources of resilient systems are exhaustive, data-centric risk assessments and risk/frequency matrixes that properly recognize and prioritize your risks. Resilient systems must be linked to an organizational culture where employees understand, embrace, support, adhere, and enforce these systems. Otherwise, they'll be ineffective and tragedies won't be prevented. Systems training must similarly be coupled with coherent checklists that support ethical and sound decision making. Such checklists will maximize employee accountability, customer loyalty, and organizational success.

STRATEGIC HINT FOR YOUR CONSIDERATION

Continuous system and employee improvement far exceeding minimum standards must be inherent in your organizational culture, as the status quo days of good enough are gone.

As you progress on our journey toward *real* risk management, the various concepts in this book will crystallize. An example is the previously discussed three precepts of *real* risk management. These precepts represent the mechanics of my philosophy and spawn all its related theories. Let's delve further into these rules. Precept one came from Archand Zeller, the great risk manager of the 1940s who codified the following assertion:

> The human does not change. During the period of recorded history, there's little evidence to indicate that man has changed in any major respect. Because man doesn't change, the kinds of errors he commits remain constant. The errors that he will make can be predicted from the errors he has made.

What does this mean? There are no new ways to get in trouble. To be sure, there are variations on errors, but in reality, it's the same type of errors over and over again. Refineries have figured out no new ways to blow up. Mines have figured out no new ways to collapse. Airplanes have figured out no new ways to crash. Ships have figured out no new ways to sink. Restaurants have figured out no new ways to poison people. And remember the following: Don't eat fish in restaurants on Monday. The fish eaten on Monday was delivered on Friday. It's four days old. What did your mother tell you about old fish? Cops, firefighters, and water/wastewater professionals have figured out no new ways to get in trouble. While I've never done your job, I've studied your tragedies: the deaths, injuries, embarrassments, indictments, lawsuits, and investigations. There are no new ways to get in trouble, and that reaffirms the importance of reviewing your past tragedies and near misses as well as those of your peers and industry trends.

STRATEGIC HINT FOR YOUR CONSIDERATION

Precept three: Identifiable risks are manageable risks, and that underscores the value of the first two components of RPM (recognition and prioritization) along with precept one.

Precept two came from my mentor, professor, and friend, Chaytor Mason, who was a risk management guru in the 1970s at the Institute of Safety & Systems Management (ISSM). Here's a capsulized version of his response when I accused him of being the smartest person who ever lived. "The smartest person in the world is the woman or man who finds the fifteenth way to hold two pieces of paper together." My instant response when I first heard this remark was confusion, but then I figured it out. While there are no new ways to get in trouble (Zeller), there are always new ways to fine tune and revisit our existing systems to prevent bad things from happening. The status quo days where we've always done it that way no longer apply. I see a lot of status quo thinking in governmental organizations, including water-related entities. There's always a better way of doing business, the fifteenth way, and we must constantly be looking for it. Your customers, personnel, and profession deserve better than minimum standards. Anything you quantify, measure, or establish a metric necessitates a constant pursuit for the next best way. There's always a better way. Continuous improvement is archetypal of high-reliability organizations and reinforces the value of systems, culture, and discipline. Achieving this type of thinking ensures alignment of employee actions and conformance with organizational mission, purpose, and systems.

Now for precept three. I've built a company out of its three words: predictable is preventable. In other words, identifiable risks are manageable risks. All three precepts form the underpinnings of *real* risk management and high-reliability organizations. When followed, your organization will distinguish itself through employee fitness, customer loyalty, and tragedy prevention. Let's spend a bit more time and refresh our knowledge of precept one: There are no new ways to get into trouble. Organizations are frequently fixated on the wrong risks. Some of the risks deemed substantial are in fact relatively benign. And some of the risks that aren't on your radar will come back to haunt you. Precept one requires perspective and diversity so you can recognize the real risks facing your organization. Such prudence requires awareness of your bias as well as a commitment to decision making by data. Adhering to this rule brings you closer to converting the five root causational factors behind all tragedies into the five pillars of success. And that's achieved through properly designed, up-to-date, and fully implemented systems. High-reliability organizations achieve this conversion by ensuring their systems are effectively mobilized through data-centric risk assessments and risk/frequency matrixes. The concept of RPM is derived from precept one and will never let your organization down as long as you broaden your outlook and expand your horizons when examining past tragedies. Now, please stay with me on this brief diversion as I highlight the problems of not spending time on risks that matter because of bias and perspective as well as lack of diversity and insufficient data.

STRATEGIC HINT FOR YOUR CONSIDERATION

Organizations are frequently fixated on the wrong risks. Some of the risks deemed substantial are in fact relatively benign, whereas some of the risks that aren't on your radar will come back to haunt you.

Whenever I'm home, my wife and I always watch Wheel of Fortune. That's our favorite show. We were watching Wheel of Fortune a while back, and the show was interrupted to bring a local news flash from Larry Welk of Action News Chopper Seven. "Larry are you there?" Yes we are, Hal. We're over Newport Beach Pier, and it has been confirmed. I repeat, it has been confirmed that there's a shark in the ocean. "Tommy, can we get a camera on that there." It's live on Action News. There's a shark in the ocean. Really? A shark in the Pacific Ocean? Who would've thought? And then the news media starts talking to uninformed people on the pier. Have you heard there's a shark in the ocean? "Heard? I saw it with my own eyes." Now what does this mean to you? "I'm never going in the water again." And why not? "It's too risky." Well, let's humorously check the data according to my unscientific projections. You're about twenty times more likely to die in a traffic collision going to the beach then getting bit by a shark. Ten times more likely to die from skin cancer by not taking care of yourself at the beach. Five times more likely to die from a dirty hypodermic needle poking your foot on the beach. Four times more likely to die from rip current. Twice as likely to die taking a selfie at the beach. What do we get worked up on? It's the sharks. We're getting worked up on the wrong stuff, the superfluous risks.

I was up in California Highway Patrol (CHP) headquarters back in the 1990s to address senior management. "Okay, Captain Graham is going to be our next presenter. Gordon you've a tendency to ramble so we're going to limit your comments to fifteen minutes. Make your point please." Okay. What's the number one injury sustained by CHP officers every year? "Well, Gordon, we know you're going to tell us; just make your point." I just made my point. I asked you what the number one injury sustained by CHP officers, and you didn't answer me. "Well, Gordon, I imagine it's back related." No it's not. "It must be neck related." No it's not. "Then what is it?" Well, back in the 1990s, when I was there, it was left knee. "Left knee?" The number one injury sustained by CHP officers back in the early 1990s was left knee. Two hundred times a day, we tell our cops to get out of the car. Now that they're out of the car, two hundred times a day, they have to do what? Get back in the car. Is there any training at the CHP Academy on how to get in and out of the car? No. "Well, Gordon, they've been doing it their entire life." Maybe the wrong way and without thirty pounds of gear hanging on their belt. That thirty pounds of gear changes a lot. And they're lurching in, and they're lurching out. They're wearing out the left knee. Go to any state trooper retirement dinner and you'll see the old guys dragging their left knee. That's an identifiable risk, and it's a risk that we missed.

STRATEGIC HINT FOR YOUR CONSIDERATION

High-reliability organizations aren't fixated on superfluous risks because they apply a data-centric approach to risk assessments and risk/frequency matrixes to ensure risks are properly recognized and prioritized.

What's the greatest risk you face? Let's take it further. How are you going to die? Here's a piece of data for you. Everybody reading this book is going to be dead someday. Here's how I hope you die on your one-hundredth birthday: You have your kids over. They bring their kids, who bring their kids, who bring their kids. You have thirty people eating at your condo, and they're giving you presents you won't use because you're one-hundred. "Grandma, it's called a hover board." Thank you. And they bring out a cake and sing happy birthday. You blow out some candles. You fall sound asleep, and you never wake up. Wouldn't that be the ideal way to go? My dad's favorite joke when he was around: "Son, when I go, I'd like to follow my grandmother who was sound asleep and not screaming like the four people in the backseat of her car." I hope you live to be one-hundred.

And by the way, medical advances coupled with lifestyle modifications may allow you to stretch your life to one-hundred. Are you modifying your lifestyle? I'm not, and I'll tell you why. The joy of being seventy requires quarterly visits to Dr. David Bloom in Newport Beach. Every three months, he's drawing blood, and one week later: "Gordon, David Bloom MD, here. Your blood results are back and, Gordon, I'm a broken record. If you knocked off the red meat, reduced the alcohol, avoided dessert with every meal, exercised a little more, reduced your stress, and slept a few more hours, then you could live to be one-hundred." And my response: Dr. Bloom based on my current lifestyle, how long am I going to make it? "Gordon, you're blessed with good genes. You'll probably make it to ninety-five." Do I want ninety-five years of doing whatever I want or one-hundred years wishing I had a steak and a drink? I'm fortunate, though, and hit the gene pool. Some of you didn't hit the gene pool and aren't going to make it to seventy. Some of you reading this book will be dead in the next five years. And some of you aren't going to naturally die. Somebody's going to kill you.

STRATEGIC HINT FOR YOUR CONSIDERATION

All three precepts form the underpinnings of real risk management, and, when followed, your organization will distinguish itself through employee fitness, customer loyalty, and tragedy prevention.

So, you're working for a water-related entity. Who's most likely to kill you? Al Qaeda, ISIS, street gangs, drunk driver? Who's most likely to kill you? Your doctor. Yes, your doctor. And before you send me your hate mail, please check the data. The number three cause of death in the United States today, right behind heart and cancer, is what? Medical malpractice. Doctors are killing a quarter-million people a year from medical malpractice. A quarter-million a year. I said that recently at a conference, and right away a hand goes up from a lady in the front row. Looking at her name tag, it referenced a major east coast university along with MD, PhD, and JD. "What you just said, sir, is incorrect. We're doing long term research at our university medical school on this issue, and what you just said is incorrect." And what does your research show? "Sir, the number is much closer to one-million than a quarter-million." Well, let's lowball it and use a quarter-million.

And here's my favorite example to underscore the degrading effects of bias and opinion on one's perspective. Several years ago, we had a police event called Ferguson that you likely heard about. One week after Ferguson, I'm at Bramshill House, which is a beautiful country estate outside of London and the previous home of the National Police Academy. The Brits regularly invite me over for a program on trends in law enforcement. So I have one-hundred UK law enforcement executives, not street cops, sitting there at eight in the morning. I'm about ready to start, and a hand goes up. I look at the guy's nametag, and he's from Wales. Fortunately, there's an Irishman next to him who could translate his question. "He wants to know what you think about Ferguson." Tell him I'm not thinking about it and neither should he. I'm waiting for the final report.

STRATEGIC HINT FOR YOUR CONSIDERATION

The concept of RPM will never let your organization down as long as you broaden your outlook and expand your horizons when examining past tragedies, near misses, and industry trends.

The Irishman wouldn't back down. "It doesn't matter what the final report says. We know the truth about you American cops. You kill too many people." Well I agree with that. "What?" You said American cops kill too many people. I'm telling you I agree. "I didn't think you would agree." Cops in America kill too many people. "How long were you a cop?" Thirty-three years. "How many did you kill?" None. "None?" Your average cop in the states goes through their entire career without killing anybody. "Now we know the truth. You Americans play games with the numbers. You kill too many people." I'm agreeing with you, but now that we've established the validity of the story, let's talk data. How many are we killing? "I don't know, but it's too many." Well, I'll give you the number, and I wrote it right down on the board in front of them: One-thousand. Every year, American cops kill one-thousand people. "One-thousand? I don't believe it. You have one-million cops in the states who are making one-million contacts a day, and they're only killing one-thousand a year. I don't believe it."

Good for you. Don't believe it. One-million contacts a day, and they're killing one-thousand people a year. My bias is now showing. Most of those people, well, they were armed violent felons. Two or three hundred are bad shootings. They didn't need to be dead right then. So on the outside, we're killing a couple of hundred people a year (which is unequivocally a tragedy), but what about medical malpractice? It's all about the data.

The preceding examples underscore the necessity of perspective and the impact of bias when evaluating risks facing your organization. It also underscores the importance of data in your decision making process. That's particularly important when recognizing and prioritizing risks facing your organization, as bias and perspective can adversely influence both activities. Equally concerning, you can't properly design resilient systems without accurately recognizing and prioritizing your risks. You must be aware of your bias and perspective to ensure your vision of risk is panoramic and not parochial.

That's a good backdrop to introduce the concept of black swans and gray rhinos. Black swans rarely occur in nature, and we can't predict when they'll be born. The risk management translations are events that have a low probability of occurring but a devastating impact upon contact. Gray rhinos are more plentiful than black swans, and nearly all rhinos are gray. The risk management parallels are threats that are highly likely to occur, yet we ignore them. And in front of every black swan, there's at least one gray rhino and likely a lot more. Human behavior draws us to black swans more than gray rhinos because of their rare sightings and devastating potential. We're fixated on the impact of black swans and less curious to uncover the root causational factors, or gray rhinos, preceding the black swan event. That equates to organizations fixating on the wrong stuff or superfluous risks. Most of your problems lying in wait are charging gray rhinos, including gray rhinos disguised as black swans. They're not black swans.

STRATEGIC HINT FOR YOUR CONSIDERATION

Bias and perspective, coupled with lack of diversity and data, can adversely influence your ability to recognize and prioritize risks, which can then lead to tragedies from black swan events that are actually gray rhinos in disguise.

That takes me to my recommended reading list. There're two great books on this concept: *The Black Swan* by Nicholas Taleb, and *The Gray Rhino* by Michelle Wucker. Let's take them one at a time. *The Black Swan* is based on what happened in the 1700s. An Englishman visited Australia, and he sees a massive bird that looks like a swan but is black. The Englishman says: "Swans are white. There are no black swans." And he asks the Australian guy: "What kind of bird is this?" The Australian responds: "That's a black swan." To the Englishman: that was an unknown, unknown. Something he hadn't even thought about, which was the possibility of a black swan. Fast-forward to September 11, 2001. One week after this national tragedy, then Secretary of Defense, Donald Rumsfeld, is on television, and the media ask him: "What's your greatest fear in this war on terror?" He responds: "My greatest fear is the black swans: the unknown, unknowns. The things we haven't even thought about." On the other hand, there are gray rhinos. Massive three-thousand-pound animals running right at us, and we refuse to get out of the way. And before a black swan event arrives, there are frequently gray rhinos running all around your organization warning of impending tragedy. Let's delve deeper into this topic, but first a brief diversion. I talk to people wherever I find myself. For my younger readers, please start talking to people you don't know. You never know who's standing in front of you or sitting next to you. The examples that follow prove that advice.

STRATEGIC HINT FOR YOUR CONSIDERATION

I always endeavor to recognize, prioritize, and mobilize an action plan for my personal gray rhinos because such preparation allows me to step away from the stampede and better prepare for an actual black swan event.

I'm flying back from Washington DC several years ago, and the guy sitting next to me seemed pretty smart. So what do you do? "I'm with the Marine Corps." And what do you do for the Marine Corps? "I'm the Commandant." I have the Commandant of the Marine Corps sitting next to me. And whenever I meet smart people, I always ask them the same type of question: What's the greatest risk you face in the Marine Corps today? This guy didn't miss a beat: "Payday loans." Payday loans? I didn't see that one coming. I thought you were going to talk about suicide, divorce, or alcohol/drug abuse because the numbers are alarming and negatively impact combat readiness. I wasn't expecting payday loans.

> Gordon, suicide, divorce, and alcohol are all issues in the Marine Corps, but when I looked for the root cause (the second this guy said root cause, I knew he gets it), when I really looked at what's wrong with the Marine Corps, it condenses down to payday loans.

And for those of you who have no clue what I'm talking about, fifteen years ago, payday loan operations were outside every Marine base. Have you ever read their small print? Four-hundred percent per annum interest rates. Once those young Marines were enrolled in such a loan, it was impossible to exit. And that then leads to suicide, divorce, and alcohol/drug abuse. To me, that was a black swan. To him, it was a gray rhino: a massive beast running right at him, and he had to do something to get out of the way.

Another airplane trip. This time, I'm sitting next to a smart woman coming back from New York and inquired what she did for a profession? "I'm a vintner and run a winery." So what's the greatest risk you face in winery operations today? "Legalized marijuana." Seriously, I thought you were going to talk about wildfires in California or boll weevil infestations in Niagara Valley. I thought you were going to talk about foreign competition. I wasn't expecting legalized marijuana.

> Mr. Graham, for fifty dollars, I can sell you one heck of a bottle of wine that'll take care of you for one night. But for fifty dollars in many states, you can buy two weeks' worth of high quality THC grade marijuana. The greatest risk we face today is legalized marijuana.

To her, that was a gray rhino. To me, that was an unknown, unknown: a black swan.

STRATEGIC HINT FOR YOUR CONSIDERATION

Black swans have a low probability of occurring, a devastating impact, and are preceded by at least one gray rhino, whereas gray rhinos are threats that are highly likely to occur, yet we ignore them.

STRATEGIC HINT FOR YOUR CONSIDERATION

The crux of real risk management is adeptly addressing and fixing root causational factors by converting these problem factors to solution factors through the five pillars of success.

And once again, I'm flying. This time to Tampa, and the passenger next to me was obviously loaded. He was wearing a watch worth about a quarter-million dollars. So how did you make your money? "Storage units." Storage units? So what's the greatest risk you face in storage units today? Are you ready? "Uber." Uber? "Yes. Pretty soon everybody's going to be using Uber. They'll own fewer cars, have open garage space, and I'll be out of business." To me that was an unknown, unknown: a black swan. To him, it was a massive beast running right at him: a gray rhino. One final story. I recently addressed a large group of sheriffs. What's the most expensive lawsuit you face, sheriff? "Well, I guess it's from jail operations." I can't speak for your department, but for your average sheriff's department around the United States, they pay a lot more money in litigation from human resources than jail operations.

Public safety organizations are sued more frequently and more severely from their own people than from inmates in their custody.

The purpose of these examples is to challenge your thinking and to assess if you see the gray rhinos impacting your organization. You have to start thinking like a risk manager. In my live programs, I always get a raised hand. "Mr. Graham, you talk about risk a lot. What's the greatest risk you face?" Oh, like I don't think about this stuff. The greatest risk I face is getting into two hundred different bathrooms every year and taking a shower. I'm always in different hotels around this great country. Taking that shower in the morning is the greatest risk I face because every hotel has a different bathroom setup. Some bathrooms have mats, while others don't. Some have shower stalls, but most have showers in the tub. Every time you enter, there's a different tub height, coefficient of friction, and slope. The first thing I do when I arrive at a new hotel is locate the emergency exit and count the steps back to my room. The second thing I do is prepare the bathroom for my shower the next morning. I always endeavor to recognize, prioritize, and mobilize an action plan for my personal gray rhinos. This level of preparation will allow me to step away from those charging gray rhinos and assist me when there's a black swan event: an unknown, unknown.

What's the greatest risks you face in your organization? How do you get hurt? How do you get sued? How do you get indicted? How do people get fired? How do we end up downstream in tragedy? There are few black swans in water/wastewater operations. There are, however, plenty of gray rhinos, which brings me to an important point. If all your executives, managers, and supervisors come from the same background and experience, then you're going to miss some gray rhinos thinking they're black swans in disguise. The solution is diversity. You must onboard people who represent the community you serve and include them in your organizational decision making process as well as your risk assessment and risk/frequency matrix evaluations. When you review risks unilaterally, sometimes you're going to miss the obvious. That's all part of the practice known as *real* risk management. What can you do up front to prevent these problems? As a lawyer, I handle tragedies. As a risk manager I prevent tragedies, and I know there's many things you can do up front to better protect yourself, your organization, your community, your profession, and our great country.

FINAL THOUGHTS

Real risk management is more than safety stuff. It's an inculcated way of thinking by high-reliability organizations to do the right thing, the right way every time. The philosophy's core precept is past tragedies are predictive of future tragedies, and, therefore, identifiable risks are manageable risks. This actuality requires continually improving resilient systems that are properly designed, kept up to date, and fully implemented. These systems must be derived from data-centric risk assessments and risk/frequency matrixes that recognize and prioritize the thousands of risks facing your organization. Such an approach is only achieved with well-informed and diverse analyses devoid of bias and opinion. Resilient systems, however, aren't enough to prevent tragedies. They must be linked to a culture of discipline where employees understand, embrace, adhere, and enforce organizational systems and their underlying rationale.

Almost all of tragedies are the result of internal errors from good employees and can be proactively prevented by identifying and addressing the root causational factors of people, policy, training, supervision, and discipline. These factors represent the problems lying in wait that produce tragedies. The crux of *real* risk management is adeptly addressing and fixing root causational factors by converting these problem factors to solution factors through five pillars of success. High-reliability organizations achieve this success and are underrepresented in tragedies as well as overrepresented in employee fitness and customer loyalty.

Next stop is the ten families of risk, which is my system for classifying the thousands of risks your organization faces into manageable families. This catalog will allow you to properly recognize and prioritize risks by applying data-centric and exhaustive risk assessments and risk/frequency matrixes to ensure your control measures (systems) reflect the actual risks arising from your problems lying in wait. Property use of the ten families will guarantee your systems are resilient and designed for those charging gray rhinos as well as any black swans that are actually gray rhinos in disguise.

Chapter Takeaway on Media Relations/Crisis Communication

Please refer to the media relations/crisis communication material in the Addendum section and ask yourself how you would address the following scenario:

Robert was a fifteen-year purchasing agent who recently left your organization. He was an exceptional employee who worked hard, was well respected, and exceeded expectations. In fact, Robert was your direct report when you were the administrative director before promoting to assistant general manager and public information officer. On Robert's last day, you wished him well but didn't ask probing questions about his exit. You now recall his fleeting comment as he left your office: "Janice, you're an empathetic, smart, and supportive leader. I wish you were still my supervisor and involved in the day-to-day operations." His comment echoes as if it's on repeat cycle. Why didn't you press him on that statement? You're also disappointed in your human resources (HR) department. They performed the requisite exit interview but didn't probe into the reasons for his departure. The HR assumed it was for a few hundred dollars a month more of pay. Robert's decision to exit was perplexing, but you let it go as a personal matter that didn't warrant your involvement or organizational action.

Several weeks later, the reasons for his departure become clear. You just concluded a phone call with a prominent investigative reporter seeking an on-the-record interview for a pending exposé she's writing on documented corruption, intimidation, and retaliation within your organization. The reporter goes on to say that Robert provided her with meticulous details of improprieties over the past three years. As she propels the allegations like water from a broken hydrant, you're unable to think clearly. How could there be such turmoil in our organization? You're responsible for organizational auditing, and you hand-picked the colleagues to perform this deliverable. Equally concerning, why didn't Robert come to you and express his concerns? Did I become unapproachable with my promotion? Did I unwittingly disconnect myself with the day-to-day operations?

You then pivot to the audits. It's true they were informal, but you appointed trusted professionals to perform this important activity. From your previous experience as a supervisor, you always felt better when audits were friendly and not overly intrusive. You then recall a board meeting last year where the governing body inquired into the prudence of hiring an outside firm to conduct an audit of the auditors. You advised against this recommendation on grounds of expense and redundancy. You're now questioning that assessment. Could there be unintended bias and insularity by you and your organization's leadership team? It's true your team lacks diversity of opinion, experience, and background, but that's only because the organization is a great place to work and promotes its leaders from within. We all collaborate and arrive together on important decisions, strategies, and findings. That's always been our culture and process.

The organization's reputation and your personal competency will soon be questioned. How could this blind spot exist in an organization with experienced management and a historical track record of low attrition? As you hang up, you know your time is limited. The investigative reporter is aggressive and respected. She most certainly has discussed these assertions with other employees. You wonder if this exposé will be used by some employees as retribution for an unpopular wage freeze implemented earlier this year. The thought of preventable gray rhinos disguised as black swans occupy your thoughts. What caused this blind spot? Could it have been our lack of diversity and opinion? It's now time to focus on the situation and formulate your solution. You must take control of the narrative, but how do you balance the need for transparency with the importance of protecting the organization's integrity and standing? What are your next moves, and how should you articulate your response?

2 Ten Families of Risk

Classifying Risks through Recognition and Prioritization

STRATEGIC HINT FOR YOUR CONSIDERATION

The ten families serve as a repository for the thousands of risks facing your organization; it provides laser guidance for recognizing and prioritizing these risks through the tools of risk assessments and risk/frequency matrixes.

DOI: 10.1201/9781003229087-2

SUMMARY

The ten families of risk is a catalog for recognizing and prioritizing the thousands of risks facing your organization. Classifying risks into select families makes the risk management process more wieldy, and that's important because identifiable risks are manageable risks.

This classification system provides laser-focus on the first two components of RPM (recognition, prioritization, and mobilization), which subsequently enables an effective final component of mobilization (action) where resilient systems are built on your root causational factors or problems lying in wait.

Chapter emphasis will center on the ten families and challenge the reader to identify and prioritize their most significant risks for each family and the necessary control measures (systems) to address their root causational factors so that tragedies are prevented and gray rhinos are managed.

Gordon Graham here, and thank you for continuing the journey toward *real* risk management. In Chapter 1, we described my philosophy as the inexorable pursuit by high-reliability organizations to do the right thing, the right way every time. This inculcated way of thinking is predicated on the continuous improvement of resilient systems that are properly designed, kept up to date, and fully implemented. Systems, however, aren't enough to achieve *real* risk management. They must be paired with a culture where employees understand, support, and enforce organizational systems and their underlying rationale.

Employees must similarly embrace the importance of conforming to prescribed systems so that activities are consistently, accurately, and universally performed. Systems and culture in high-reliability organizations are self-reinforcing, with discipline serving as the adhesion. This linkage correlates employee behavior to organizational mission and purpose while concurrently serving as a governing luminance for ethical and sound decision making.

An important derivative of *real* risk management is the conversion of problems lying in wait, known as root causational factors, into pillars of success. Good people who act with good policy, who are regularly trained, properly supervised, and in an organization that addresses arrogance, ignorance, and complacency with fair and impartial discipline, will prevent tragedies by doing the right thing, the right way every time. The alternative, however, is also true, as root causational factors are the precursor to tragedies if not identified and addressed. Figure 2.1 is a reproduction of Figure 1.1 and a reminder that root causational factors straddle the line between pre-incident prevention and post-incident correction. High-reliability organizations understand this duality and proactively transform their problem factors to solution factors. The outcome is an underrepresentation in tragedies and an overrepresentation in employee fitness and customer loyalty.

Chapter 2 is a culmination of my studying and handling tragedies as a risk manager and lawyer for forty-plus years. Early on, people were coming up to me and saying: "Gordon, I like what you're espousing, but where do we get started? We face thousands of risks as a public safety organization." And I told them I don't know.

Organizational Risk Management

High-reliability organizations build resilient systems that are properly designed, kept up-to-date, and fully implemented. These systems proactively prevent tragedies by addressing its derivation, which are the root causational factors. Successful systems require recognizing and prioritizing risks through ongoing and data-centric risk assessments and risk/frequency matrixes as well as embedding an organizational culture of employee discipline, continuous improvement, and system conformance.

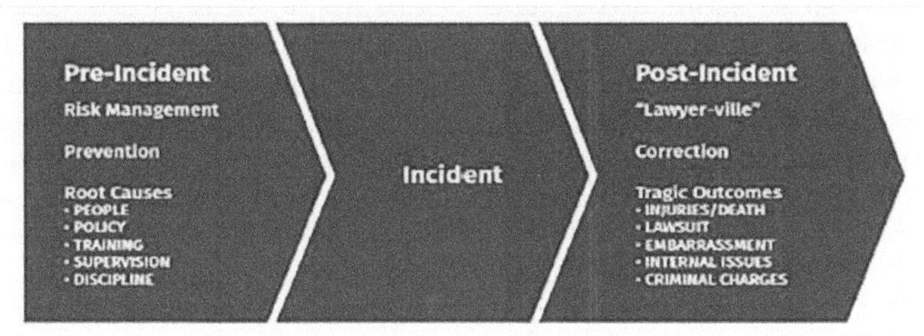

FIGURE 2.1 Organizational risk management. Source: Institute of Safety & Systems Management (ISSM).

The question kept surfacing so I put together a series of sequential programs that naturally build upon each other and allow organizations in all professions, including water/wastewater, to understand and implement *real* risk management. The first of these programs is entitled the ten families of risk and is based on my belief that pre-incident prevention is operationally achievable as well as more affordable and less detrimental than post-incident correction. Its effectiveness requires proactive management of your root causational factors, and that necessitates an exhaustive and data-centric approach for recognizing and prioritizing the thousands of risks facing your organization. The investigative tools to perform the first two components of RPM (recognition, prioritization, and mobilization) are risk assessments and risk/frequency matrixes. The ten families envelop these tools and provide a workable catalog as well as directional focus and laser guidance for those risks that pose the greatest danger as opposed to organizational fixation on risks that are superfluous and not deemed serious or plausible.

STRATEGIC HINT FOR YOUR CONSIDERATION

The proper function of the ten families as a comprehensive risk catalog requires its investigative tools be devoid of bias and opinion as well as contemplative of diverse organizational perspectives and inclusive decision making.

The volume and assortment of risks facing water-related entities compel a coherent classification mechanism. The ten families, like all concepts derived from *real* risk management, is predicated on three fundamental precepts: (1) past tragedies are predictive of future tragedies; (2) there are always better ways to prevent tragedies through system improvement; and (3) predictable is preventable, meaning identifiable risks are manageable risks. These precepts yield several overlapping theories, including RPM. The ten families is a repository for the first two components of RPM and encompasses the investigative tools of risk assessments and risk/frequency matrixes. The former are analytical devices to detect and classify risks within the ten families, whereas the latter rank these risks in each family according to probable frequency, severity, and time to think (non-discretionary or discretionary time). Throughout the book, we'll refer to activities that are high risk, low frequency, with no time to think (non-discretionary time) as core critical tasks and activities that are high risk, low frequency, with time to think (discretionary time) as critical tasks. Most employee errors are caused by core critical tasks and necessitate daily training. Critical tasks are a secondary source of errors and also require frequent training.

The proper function of the ten families as a comprehensive risk catalog requires its investigative tools be devoid of bias and opinion as well as contemplative of diverse organizational perspectives and inclusive decision making. The final component of RPM is mobilization or action, which serves as enterprise kinetic energy through the deployment of resilient systems. You can't design resilient systems without first recognizing and prioritizing organizational risks through data-centric risk assessments and risk/frequency matrixes. That's important because the recognition component of RPM is based on precept number two (past tragedies are predictive of future tragedies), and its effectiveness is conditioned on properly reviewing your past tragedies and near misses as well as the tragedies of your peers and industry trends of similar organizations. The ten families provide the necessary laser guidance and consolidated repository for high-reliability organizations to properly catalog and sequence their risks. Figure 2.2 details the family of risk categories and serves as the encampment for our examination.

Before we begin our review of each family, let's discuss the order. Lists are invariably prioritized according to what is deemed important, and that also applies to the ten families of risk. I don't know the religions of my readership, but I'd imagine a lot of you identify as Christians. I'm Catholic by upbringing, but there are many other denominations such as Presbyterian, Lutheran, Methodist, Anglicans, and Episcopalian. Nondenominational Christians are also under this grouping. There are several excellent and equally fulfilling religions, but let's talk about mine for a moment. Catholicism has the Ten Commandments. I first learned about the Ten Commandments in the 1950s

Ten Families of Risk	
➲ External	➲ Legal & Regulatory
➲ Strategic	➲ Organizational
➲ Operational	➲ Informational
➲ Technology	➲ Human Resources
➲ Financial & Reputational	➲ Political

FIGURE 2.2 Ten families of risk. Source: Graham Research Consultants, LLC.

at a Catholic grade school. I had a question. How did they establish the order for the Ten Commandments? Did they throw them in a hat and pull them out at random, or is there a reason the First Commandment is the First Commandment? There's a reason for *Thou Shalt Not Have Any God Before Me* being number one. When you write a list, you prioritize it based on what you deem is important.

Maybe there are some constitutional scholars reading this chapter. What was going on in 1787? We had Ben, Tom, and Betsy standing around one night having a Sam Adams and getting angry. Betsy was irate. "You know, the British soldiers kept on kicking in my door and made me feed them and give them a bed. That's not right." Tom was fuming. "They kept on trying to take away my gun. That's not right." Ben was outraged. "They kept on breaking up my printing press." That's why when the founding fathers forgot to put some protections in our Constitution, they came up with the first ten amendments. How did they establish the order for the ten amendments? Did they throw them in a hat and pull them out at random, or is there a reason the first amendment of the Bill of Rights (freedom of speech, religion, press, and assembly) is number one? When you write a list, you prioritize it based on what you deem is important. That's what I've done with my list. Let's now examine my ten families of risk in detail.

FAMILY ONE

External

When I formulated the ten families of risk, I chose external as number one for a reason: I like to complete the most complicated activities first. The most difficult family of risk you face is number one, external. Unlike families two through ten, we have no control over these risks. External risks arise outside your organization and impact your operations. They're increasing in frequency and severity and pose a complacency exposure for organizations, especially with disaster planning and terrorism. Pandemics, drought, terrorism, and organized crime are external risks. If your building is in the flight path of an airport, that's an external risk. If you're adjacent to a major body of water, that's an external risk. If you have pipelines underneath your building, that's an external risk. If you have a prison in your community, that's an external risk. If you have train tracks going through your community, that's an external risk. If you have an interstate going through your community, that's an external risk.

STRATEGIC HINT FOR YOUR CONSIDERATION

The ten families provide the necessary laser guidance and consolidated repository for high-reliability organizations to properly catalog and sequence their risks.

STRATEGIC HINT FOR YOUR CONSIDERATION

Your disaster management plans should be current, consistent, and aligned with all impacted stakeholders so that when something bad happens, everyone affected is synchronized in their actions.

I live in Huntington Beach, which is south of Seal Beach in Southern California. There's an underground weapons repository in Seal Beach for the Pacific Fleet. So we have bombs in Seal Beach. How do they get there? Train tracks, which run right into the weapons yard. About fifteen years ago, I was listening to the scanner for Seal Beach Police Department: "We have an overturned train going into the weapons yard." Thirty seconds later, dispatch comes back and says what? "The twenty-four hour number we have on record is disconnected." Now, it ended up being an area code issue, but even an area code issue can be problematic. "Gordon, that was several years ago. Anything more recent?" Yes. A few years back, cops were chasing a suspect in the middle of Central California. The pursued was heading to a nearby naval air station where they train fighter pilots to land on carrier decks. The cops radioed dispatch to have the naval air station close their front gate, but the twenty-four hour telephone number on record was disconnected. Shortly thereafter, the suspect blew through the entrance barrier and collided with an F-18 fighter jet. A fifty-million dollar loss, all for a bad phone number.

The current thinking of many organizations is that disasters won't happen, and if they do occur, the event won't be impactful. And should a severe disaster occur, then there wasn't anything the organization could've done up front to prevent it. Let's examine that data point further. For too many professions, the disaster management plans are out of date. You'd assume multi-billion-dollar organizations have well-written disaster management plans codifying what to do in the event of a particular tragedy from their operations. They don't. Many of their plans are seriously out of date. I'm still troubled with what's going on in Florida. We had the Parkland shooting, and there's still disagreement between firefighters and cops. The former want children to evacuate the building when an alarm goes off, whereas the latter want children to shelter in place to prevent a chokepoint at the exits. Even firefighters and cops have disagreements on disaster management plans, and they're more up-to-speed than most professions.

STRATEGIC HINT FOR YOUR CONSIDERATION

Mike Tyson famously said "everyone has a plan until they get punched in the mouth," meaning the key to planning is training as well as plan adaptation and flexibility to address natural contingencies that always surface.

On my recommended reading list is a great book by Andrew Hopkins titled *Disastrous Decisions*, and it's all about the Deep Horizon spill. Remember that one? Gulf of Mexico and the well blew up. British Petroleum (BP) was the owner, and they have wells throughout the Gulf and in various other bodies of water. BP knew a blowout would eventually occur and put together a control measure (system) of what to do if that scenario transpired. Their disaster management plan was well written and several hundred pages long, including four pages on protecting walruses in the event of an oil spill. How many walruses make their home in the Gulf of Mexico? None. BP got caught. Where did they generate the disaster management plan for their multi-billion-dollar project in the Gulf of Mexico? They lifted it from an out-of-date plan for their oil wells off the coast of Aberdeen, Scotland. All they did was perform a find/replace search function on their word document and changed North Sea to Gulf of Mexico. Two-thousand phrases were changed. You should read *Disastrous Decisions.*

Disaster management plans must be taken seriously. So here are my serious questions for you and your organization: Are your disaster management plans up to date? Are they complementary to the disaster management plans of police, fire, and other first responders that will be assisting with your tragedy? What about other major industries in your community? Your disaster management plans should be current, consistent, and aligned with all impacted stakeholders so that when something bad happens, everyone affected is synchronized in their actions. Equally important, your personnel should know the employees of those organizations assisting in the tragedy. The mere fact that someone knows your first name will help facilitate a better response to a tragedy in your community. Mike Tyson famously said "everyone has a plan until they get punched in the mouth," meaning the key to planning is your counterpunch or reaction after that punch to the mouth. Make sure you've planned appropriately and are flexible in adapting your plans for the natural contingencies that always surface. In my forty-plus years of responding to wildfires, earthquakes, riots, plane crashes, and similar tragedies, I've witnessed a lot of great plans that fell apart after a punch to the mouth. To prevent this unraveling, your plans must be continuously reviewed, properly designed, and up to date. Training is essential to ensure your personnel execute their specific role in real time and under pressure.

STRATEGIC HINT FOR YOUR CONSIDERATION

Many water-related entities aren't in compliance with the laws and regulations that control their existence. If you aren't in full compliance with federal and state laws, and someone suffers a harm or loss, it's instant liability.

The most difficult external risk to prevent is terrorism. Your best mitigation measures for this type of misconduct are vigilance and random irregularity. These actions will maximize your ability to thwart any deliberate, willful, and malevolent behavior. You must be vigilant. Sadly, vigilance wanes over time and complacency becomes the norm until a terrorist act occurs, and then the vigilant sequence repeats itself. Another great book is *Predictable Surprises* by Bazerman and Watkins. It details the cycle of disaster and complacency. Next, we need to instill the value of random irregularity. Your standardized practices and operational schedules shouldn't be too predictable. Bad actors study their intended targets and look for patterns. They then exploit those patterns to help achieve their nefarious goal.

FAMILY TWO

LEGAL AND REGULATORY

Family two is legal and regulatory risks. This family arises from the complexity of, or non-compliance with, the legal framework imposed on water-related entities. If there's a law or ordinance in state or federal statutes that requires action or prohibits action, this rule must be known and followed. Are you in full compliance with the laws and regulations that control your existence? And before you say, yes, how do you know? While I've never done your job, I've studied your tragedies. All too often, lawyers performing post-incident investigations identify the involved water-related entity as not being in compliance with laws and regulations. That's a problem lying in wait so please talk to your competent counsel to make sure your organization is in compliance with all appropriate laws and regulations. It requires properly designed, up-to-date, and fully implemented policies (systems) that are regularly reviewed.

Are you in full compliance with the alphabet soup of regulations including the Americans with Disabilities Act (ADA), Fair Labor Standards Act (FLSA), Equal Employment Opportunity Commission (EEOC), Health Insurance Portability and Accountability Act (HIPPA), Family Medical Leave Act (FMLA), and Fair Credit Reporting Act (FCRA)? There are many more, and each of these laws and regulations has numerous subsections that require or prohibit actions on behalf of your employees and public. I'm always amazed by how many organizations aren't in compliance with the laws and regulations that control their existence. Have you enacted all the policies (systems) required by state and federal laws and regulations? One example would be federal laws to accommodate people with hearing disabilities, sight disabilities, and limited English proficiency. If you aren't in full compliance with federal and state laws, and someone suffers a harm or loss, it's instant liability. Are you in compliance with the training standards (hours and frequency) required by your state for licensed water operators? How about your state's public records act?

STRATEGIC HINT FOR YOUR CONSIDERATION

Strategic risks require looking into the future and identifying emerging issues such as regionalization, consolidation, climate change, technology, and outsourcing that may cause your organization to be redundant or unnecessary.

FAMILY THREE

Strategic Risks

Family three, strategic risks, is one of my annoyances in life. You must look into the future. I'm a patriot to our great country, but for your average American, the future is Friday. The distant future is next Friday. Maybe you saw the military leader of China give a three-hour monologue a few years ago. You can receive an English translation by performing an internet search. You should read the entire monologue. He's not talking weeks, years, or decades. He's talking centuries. And we're worried about Friday. Strategic risks arise from the lack of priority setting and business planning that lead to a reactive organization not being prepared or equipped to deal with unforeseen events. This torpor is especially evident in smaller organizations that don't have sufficient personnel or time to plan for the future. Where will your organization be in thirty years? You must think strategically. Let me accentuate the point. The Wall Street Journal recently had a story where forty percent of Americans couldn't come up with four-hundred dollars within twenty-four hours of an emergency. Your average American aged fifty has less than twenty-five-thousand dollars saved for retirement outside the value of their house. Your average American aged twenty-five has nothing saved for retirement. "Well, Gordon, we have pensions." They're upside down throughout the United States. "Gordon, we have social security." With our declining birthrate, social security is going to be more stressed than it is today. Prepare for the future, and always remember it begins at the individual level. If you aren't planning at home, then you're likely not planning at the organization.

And for the executives, don't assume your organization or profession is immune from change. It isn't. Regionalization, consolidation, outsourcing, technology, population change, and climate change will affect water-related entities. The innovative executives understand this projection and will prevail. Those that don't, won't. On my recommended reading list are two great books by George Friedman: *The Next Hundred Years* and *The Next Decade*. Read them in that order. They make more sense. These books are difficult to read, as Mr. Friedman is a complex writer. You want to read a simpler book? Spencer Johnson has one called *Who Moved My Cheese*. Smart people are looking into the future with this question: Where's the cheese going to be, and what do I need to be doing today to make sure I receive a piece of that cheese in the future? I regularly tell young adults that employers will be pursuing them if they're literate in computers and Mandarin. The same applies for diesel mechanics bilingual in English and Spanish. I had my own early experience with strategic risk management. In the 1960s, my mother enrolled me in a typing class because she saw the value in developing that skill. Strategic risks require looking into the future and identifying the risks that may cause your organization to be redundant or unnecessary.

STRATEGIC HINT FOR YOUR CONSIDERATION

Too many of your tragedies come from your own good people making mistakes, errors, lapses, and omissions in the area of core critical tasks and critical tasks.

FAMILY FOUR

ORGANIZATIONAL RISKS

Family four, organizational risks, requires a universal methodology to managing the overall risks of your enterprise. This family comprises wide-ranging risks arising less from operational activities (that's family five) and more from the failure to clearly define employee roles and responsibilities as well as not adhering to organizational values and not having monitoring processes (feedback loops) in place. You'll see much more detail on organizational risk management in subsequent chapters, but let's review certain recurring themes that transcend proximate cause and are the real precursor to tragedies. These themes involve problems lying in wait or root causational factors. The derivation of most, if not all, tragedies is lack of quality people, policy, training, supervision, and discipline.

Let's briefly examine these root causational factors. The principal components for getting and keeping good people are recruitment, backgrounds, probation, and performance evaluations. I'll assume you have good people, but good people aren't enough. I'm concerned about the quality of your control measures, policies, procedures, processes, rules, checklists, and protocols (systems). Are they properly designed, up-to-date, fully implemented, and taken seriously? This unease stems from my review of policy manuals from water-related entities throughout the United States. I regularly see organizations with no policies or policies that are out of date, inconsistent, antiquated, and wrong. I'm asking you to review your policy manual to make sure your policies (systems) are properly designed and kept up to date.

STRATEGIC HINT FOR YOUR CONSIDERATION

Organizational Risk Management involves problems lying in wait or root causational factors. The derivation of most, if not all, tragedies is lack of quality people, policy, training, supervision, and discipline.

My major concern is training or pillar three. I believe you have good people. I believe you have good policy or can get good policy. Training distresses me. Just a quick thought. After you were hired as a full-time employee and released from probation, when was the next time you took a serious test involving study preparation? Probably never if you chose not to promote. Well, I'm not picking on you. Public safety professionals who are released from probation and choose not to promote don't have to take another serious test involving advanced preparation. When I speak to these groups, I always follow with this question: Do you get your hair done professionally? In many states, the man who does your hair is required to take a regular test to make sure he has the necessary skills and abilities to cut hair. We have cops carrying guns with bullets who haven't been tested on shooting policy since point of hire, and we have firefighters driving fifty-thousand pound apparatus who haven't been tested on vehicle operations policy since point of hire. That's a problem lying in wait.

Back to water-related entities. You have in your organization what I refer to as core critical tasks (very risky, done very rarely, with no time to think) and critical tasks (very risky, done very rarely, with time to think). Every job description has a small number of core critical tasks (non-discretionary time) and a larger number of critical tasks (discretionary time). Core critical tasks are overrepresented in tragedy and critical tasks are right behind them. Of the thousands of activities for a particular job description, there's a small number, less than fifteen, that are overrepresented in tragedy. These core critical tasks require constant training. It's important to make every day a training day and focus the training on core critical tasks. In the public safety business, shoot–don't shoot and two-in-two-out are core critical tasks. In the water/wastewater business, a chemical spill is a core critical task (very risky, done very rarely, with no time to think). Heart attack in the workplace and the proper use of an automated external defibrillator is a core critical task. Workplace violence and building evacuations are core critical tasks. I can't overemphasize the following: Good people I believe you have. Good policy I know you can get. But you have to make every day a training day and focus your training on core critical tasks (very risky, done very rarely, with no time to think or non-discretionary time) and critical tasks (very risky, done very rarely, with time to think or discretionary time).

STRATEGIC HINT FOR YOUR CONSIDERATION

When people don't follow policy (systems), it must be addressed with discipline, notwithstanding outcome. In too many organizations, discipline is a function of consequence.

The fourth of the five pillars of success is supervision. Chapter 7 is dedicated to supervision, but let's give a primer on the topic. You show me a tragedy in a water-related entity, and I'll show you a proximate cause of X. The real problem lying in wait, however, is all too often a supervisor not behaving like a supervisor, or a supervisor who tried to behave like a supervisor but wasn't supported by management. If either is present, that's a problem lying in wait. And the fifth of the five pillars of success is discipline. When people suffer from arrogance, ignorance, or complacency, there must be discipline. When people don't follow policy (systems), it must be addressed with discipline, notwithstanding outcome. In too many organizations, discipline is a function of consequence. "How did it end up?" Okay. "Well, all's well that ends well." Discipline is never a function of consequence. It must be a function of policy. When people don't follow rules, it must be addressed. That's family four, which comprises getting and keeping good people, building good policy, making sure your people are fully and adequately trained on core critical tasks and critical tasks, assuring supervisors are behaving like supervisors, and disciplining when rules aren't being followed.

FAMILY FIVE

Operational Risks

Family five is operational risks, which is best defined as problems arising from specific activities (tasks, incidents, and events) performed by your employees in every job description. Unlike organizational risks, which are universal and wide ranging, operational risks are tactical and linear. This family correlates with the inexorable pursuit by high-reliability organizations of doing the right thing, the right way every time. And that goes back to specific activities (tasks, incidents, and events). The vast majority of activities are performed correctly. That's the good news. The bad news is occasionally things don't go right. When that happens, you end up on the right-hand side of Figure 2.1, which we call Lawyerville. The right side is bad. When things don't go right, people can die or be injured, including our employees. You can also be embarrassed, indicted, fired, or sued. All of those outcomes are the domain of lawyers. What can you do up front to assure things go right and avoid Lawyerville? You must build better control measures (systems), and all personnel must be adequately trained to correctly perform their core critical tasks (non-discretionary time) as well as understand the value of thinking things through on critical tasks (discretionary time) when time allows. Let's examine this question more completely.

STRATEGIC HINT FOR YOUR CONSIDERATION

Family five is operational risks, which is best defined as problems arising from specific tactical and linear activities (tasks, incidents, and events) performed by your employees in every job description.

How do you manage the risk of a specific activity like employee termination or pathogen incident? It begins with RPM (recognition, prioritization, and mobilization) and ends with daily training for high-risk, low-frequency activities where you have no time to think (core critical tasks) as well as high-risk, low-frequency activities where you have time to think (critical tasks). The good news is most of what your personnel do, they're doing right. Why? Most of what your people do, they've done before. Your brain is an accumulation of past experiences, memory markers, behavioral scripts, and mental models. When the brain performs an activity, the brain scans past behavior. Have I done it before? Lots of times. Did it end up going right? Every time. So rather than thinking it through, why don't I just do it the same way as last time? I bet it goes right, and if it goes right, I stay out of trouble. This theory is known as recognition-primed decision making (RPD), and we'll delve into its details in Chapter 4 along with Gary Klein's great book *Sources of Power*.

FIGURE 2.3 Tragedies: Types and sources. Source: Graham Research Consultants, LLC.

We just discussed the reason most activities go right. What about when they go wrong and tragedy occurs? There are two types and sources of tragedies: (1) somebody did something bad on purpose, or (2) somebody made a mistake. Both can be caused from internal or external sources. Figure 2.3 illustrates these sources and types of tragedies. Internal intentional misconduct is addressed when our people component is in place. In Chapter 5, we'll talk about the importance of comprehensive background investigations. External intentional misconduct is difficult to prevent and requires vigilance and random irregularity (family one). Negligent conduct is attributed to good people from inside or outside your organization making mistakes, errors, lapses, or omissions. The majority of this negligent conduct is attributed to your employees, and, therefore, can be proactively managed. Mistakes don't impulsively surface. They're usually in that area of core critical tasks and critical tasks: very risky, done very rarely, with no time to think (non-discretionary time), or very risky, done very rarely, with time to think (discretionary time). Again, as a refresher, I must emphasize the importance of RPM (recognition, prioritization, and mobilization) and training. You must have viable policies and procedures (systems) in place to assure your employees do the right thing, the right way every time, and these systems must be built on the real precursor of tragedies, known as root causational factors or problems lying in wait. Employees must be actively trained on your systems with an emphasis on core critical tasks and critical tasks.

STRATEGIC HINT FOR YOUR CONSIDERATION

There are two types and sources of tragedies: (1) somebody did something bad on purpose, or (2) somebody made a mistake. Both can be caused from internal or external sources.

Now, it's time to crystallize the importance of training by using core critical tasks as an example. This evening when you get home, and your twelve-year-old girl or boy is standing there. "Hey Mom? Hey Dad?" "Yes sweetheart. How can I help you?" "I have a report for school that's due tomorrow." And as the parent, what's the first question you ask? "So when did you get this assignment?" "Three months ago." Three months ago? There's your child putting it off to the last minute. If I were in charge, I'd have every student write a report on Richard Rescorla. You know my bias. Every American child should write a report on this hero. There's a reason I'm telling you this story. "Well, Gordon, who's Richard Rescorla?" He's a famous American, born oddly enough in Cornwall, England. By the age of seventeen, he demonstrated tremendous athletic skills and thinking power. If you have a child in the United States who's a scholar-athlete, the coaches are calling universities. That's not the way it works in England. When there's a sharp child who's also athletic, the coaches are calling the military. And the British military came out and took one look at young Richard Rescorla and said: "The Queen needs your service."

When you research Richard, read his history in the British military. It's impressive. He ended up being an American, and we sent him to Vietnam. You should also read his history in Vietnam. Richard was highly decorated and enrolled in the GI Bill where he received a degree in emergency management. Richard subsequently was hired by a bank, and, several acquisitions later, he became vice president of emergency management for Morgan Stanley. Richard worked at their New York City headquarters and got promoted to risk manager in 1987. The first thing he did as the risk manager was perform a risk assessment. How could I do great damage to our headquarters building? Read the story. He brought over friends from the British military, and they took a look at the World Trade Center and said: "The underground parking has no gate. There are no control measures. People can come and go as they want. The underground parking was the weak link." So Mr. Rescorla told his management of the building's vulnerabilities. Management had all sorts of excuses why they couldn't break the lease. It's very expensive. It'll take too much time to find replacement space. We can't do it. And for those of you really into history, terrorists originally tried to blow up the World Trade Center in 1993 from the garage. Richard was right: The garage was the weak link.

STRATEGIC HINT FOR YOUR CONSIDERATION

Operational risks require viable systems to ensure activities are performed accurately, and that necessitates risk recognition and prioritization so that your root causational factors are converted to pillars of success.

After they arrested the blind Sheikh and his coconspirators in 1993, Richard said they're going to be back, and they're going to do the job better next time. We need to break our lease, and management again refused and reiterated the same reasoning: "We've a fifty year lease. It'll cost millions to break it." No one would listen, so he did the next best thing, and there's a reason I'm telling you this story: He started doing emergency building evacuations for all Morgan Stanley employees. Okay, folks. Everybody here on the fifty-first floor needs to evacuate. "But I'm on a long-distance phone call."

Well how about that? They just hung up. And everybody headed for where? The elevators. Don't use the elevators. Use the stairs. For most people who work in a high-rise, they've never seen the stairs. "We have to walk down the stairs, but we're on the fifty-first floor?" That's why we need to start walking now. And some of the women said, "I have high heels on." And his response: Why don't you get a pair of tennis shoes and put them in your bottom desk drawer for the just-in-case file. That one hint, storing tennis shoes, paid off a thousand times. When they were used to walking down the stairs during these drills, that's when Richard turned off the lights. "We can't see." Cell phones had to light the way. When they were used to that, he started blocking the stairs with debris. Everybody disliked him. When they had to evacuate on September 11, 2001, they had memory markers for exiting the building. They all got out safely. After a headcount, Richard went back to rescue other people, and that's where he got buried. It's an inspirational story about real heroism for every child to know and write about. Back to my point: Richard made every day a training day.

I mentioned that I fly quite often. Five minutes before takeoff, there's a charming young man or woman giving a preflight safety briefing. What's he or she training us on? The evacuation in the event of an emergency on the takeoff roll. To me, that's a core critical task: very risky, done very rarely, with no time to think (nondiscretionary time). I pay attention, but for most people on the airplane, they aren't paying attention. They're reading, working, or playing games. This core critical task isn't on their radar as a personal problem lying in wait. That makes the flight attendants upset. I'm far from perfect but regularly flight attendants come up to me and say: "Thank you for paying attention, sir." They know who's paying attention. Every day must be a training day because every day comprises operational risks that may involve core critical tasks or critical tasks.

STRATEGIC HINT FOR YOUR CONSIDERATION

Information risks arise from untimely, inaccurate, or unreliable information that influences the discharge of roles and responsibilities and are exacerbated by the dangers of ignorance, complacency, and cognitive bias.

FAMILY SIX

Information Risks

Family six is information risks, which arises from untimely, inaccurate, or unreliable information that influences the discharge of roles and responsibilities. You must be aware of the dangers arising from ignorance, complacency, and cognitive bias. There similarly must be a free flow of information up and down the chain of command. We make decisions based on information. How do you know the information on which you're basing decisions is accurate and not manipulated by someone for his own advantage? How do you separate valuable information from raw data? People know how to influence you. On my recommended reading list is a great book by Robert Cialdini called *Influence*. There are techniques people use to influence your thought process and decision making. I'm leading up to something. In too many organizations, employees are taught how to perform operational activities as opposed to how to think. "Gordon, why should we care? By your own admission, most of what we do, we do right." That's because most of what we do, we've done before; our recognition-primed decision making (RPD) is activated. What happens when we put our people in a low-frequency event? Something they haven't experienced before. Have we given them a systematic decision making process on how to think things through? "Gordon, do you have a decision making process?" I do, and we'll devote an entire chapter to ethical and sound decision making. I strongly recommend training all personnel on deliberative thinking skills, particularly on specific activities in a given job that have the highest probability of ending up in tragedy (i.e., core critical tasks and critical tasks).

FAMILY SEVEN

Technology Risks

Family seven is technology risks. I previously identified family one, external, as the most difficult family we face, but technology is the fastest growing family. These risks arise from outdated or unreliable information systems where user requirements aren't met. I have major concerns with this family, including external and internal security measures, hacking, malware, social media, compatibility of systems, lack of qualified chief technology officers (CTO), and many others. Remember this apropos quote: "In the land of the blind, the one-eyed man is king." Just because someone knows more than everybody else, doesn't make her a CTO. That's a problem lying in wait.

STRATEGIC HINT FOR YOUR CONSIDERATION

Family seven, technology risks, is the fastest growing family. These risks arise from outdated or unreliable information systems where user requirements aren't met.

FAMILY EIGHT

Human Resources (HR)

Family eight, human resources (HR), is the most expensive family of risks and comprises a myriad of issues involving personnel. It's also the easiest risk to address.

Executives, managers, and supervisors shouldn't make employment law decisions without first consulting with competent counsel or HR personnel. The reason is simple: We don't understand all of the issues involved in employment law. It's legally complex and continually changing. Transfer the risk to someone who understands the issues and performs these activities at the highest frequency. Let's reverse the scenario. You're experiencing a terrorist event where somebody barricades himself in your chemical storage room. You have police, fire, and FBI in your office. What's this guy going to do? What are we going to do?

During the deliberation, an employee advises that HR is on the outer perimeter and has advice on handling the situation. Think that one through. If HR had advice for you on an incident like this one, would you think they were obtuse? That's exactly what HR is saying when you think you understand employment law. Your great fear should be an executive, manager, or supervisor making an employment law decision, and it ends up downstream in litigation. Even if you win, you'll still pay a fortune. There are rarely any HR issues that must be decided on non-discretionary time. Slow down and talk to competent counsel or HR personnel before you make a decision that will come back to haunt you. All HR issues must be taken seriously and require assiduous contemplation. It's a critical task, meaning it's high risk, low frequency, with time to think.

STRATEGIC HINT FOR YOUR CONSIDERATION

Employment law, which is complex, ambiguous, and continually changing, is almost always discretionary time and necessitates its transfer from executives, managers, and supervisors to a competent counsel and human resources personnel.

STRATEGIC HINT FOR YOUR CONSIDERATION

If a social media lie gets repeated three times, it's perceived to be true. Your best mitigation measure is an active monitoring and retort presence where you can quickly respond and correct content that's inaccurate.

FAMILY NINE

Financial and Reputational Risks

Family nine is financial and reputational. Let's begin with financial risks. Anything dealing with money is filled with peril. Financial risks comprise improper budgeting, forecasting, and expenditure controls as well as contracting, asset management, internal audits, improper salaries, overtime misuse, resource misappropriation, embezzlement, and poor revenue oversight. I'm always amazed at who's managing (or not managing) the money. Is your chief financial officer qualified for the position? Does she have a background in finance, or did she test well on promotional examinations? Let's assume two of your employees are communicating about overtime. "Mary, I see you're working next Thursday. I'm working next Friday. Why don't you call in sick on Thursday, and I'll backfill you on overtime. I'll then call in sick on Friday, and you backfill me." That's a criminal conspiracy to defraud organizational resources. It's a crime, and it's systemic. Budgets are filled with risk. Special project money is filled with risk. Time for another diversion to accentuate the point on the importance of operational vigilance when dealing with finances.

So, I finally asked out the woman in the bank. I've been building up the courage for a couple of years. I was a brand new lawyer, brand new sergeant, and feeling lucky. My secret crush was counting out my money for the month and said: "Will there be anything else?" So, I finally did it. I'd like to take you on a date, and she shocked me when she said, "I'd love to." We ended up getting married and are still madly in love. That was back in 1982. Three years later, we're watching Wheel of Fortune, and the prize puzzle was a trip to Germany. My lovely wife says three years into our marriage: "I'd love to go to Germany." And I responded that I'll take her to Germany. Her retort: "You never take vacation. We've been married for three years, and you don't take vacation." Well, sweetheart, I'm banking my six-hundred-fifty. "What does that mean?" Sweetheart, at the California Highway Patrol, you can save up to six-hundred-fifty hours of vacation time, and when you max-out your bank, you can start burning your annual leave, but you always keep your bank at six-hundred-fifty in case of catastrophic illness. At the end of your career, you can sell it back to the state at a price greater than when you earned it. That's why I'm banking my six-hundred-fifty.

STRATEGIC HINT FOR YOUR CONSIDERATION

Successful systems are built on root causational factors, known as problems lying in wait, and these factors must be identified and addressed before their conversion into pillars of success.

Her response to that was: "You can't do that in the banking business. I have to take two weeks consecutive vacation every year. If you come on bank property during the two weeks, then you're instantly terminated." Why? "If you're pulling a financial scam, it'll unwind in two weeks." To me, that was a black swan. To her, it was a gray rhino. Why am I telling you this story? Five or six years later in Dixon, Illinois, a well-liked woman who never took a day off had been stealing money from the city for years. Fifty-three-million dollars was the total, and she never took a day off. Make sure you have control measures (systems) with respect to expenditures, budget, credit cards, petty cash, and vacations.

Let's talk about your reputational risks. It isn't uncommon for people to speak poorly about governmental organizations, including water-related entities, on social media. Here's an unfortunate rule: If the lie gets repeated three times, it's perceived to be true. Your best mitigation measure is an active monitoring and retort presence on social media where you can quickly respond and correct content that's inaccurate.

FAMILY TEN

Political Risks

Family ten is political risks. Dealing with publicly elected officials is filled with risk so please be careful and remember it's all discretionary time. All personnel should refrain from the theater of politics. There's a lot of risk involved here, and your organization's mission shouldn't be impacted by which political party is in control of your state. This family is a tough one, particularly if your organizational governing body is elected officials. If you think politics are tough at the federal or state level, take a look at what's going on in your local government.

FINAL THOUGHTS

The profession of water-related entities is inherently dangerous. It involves thousands of activities (tasks, incidents, and events) and subsequently correlates to an equal number of risks. The cataloging of these risks into ten families makes the process more manageable, which is important because identifiable risks are manageable risks. The ten families is augmented by the first two components of RPM, recognizing and prioritizing risks. The former includes data-centric risk assessments based on your past tragedies and near misses, as well as those of your peers and industry loss trends for similar organizations. The latter ranks your risks according to probable frequency, severity, and time to think. The ten families amalgamates these components and provides necessary directional support and repository facilities, which then enables an effective mobilization (action) component through resilient systems. Successful systems are built on root causational factors, known as problems lying in wait. These factors must be identified and addressed before their conversion into pillars of success. The ten families provides this capability through its investigative tools and similarly ensures proper awareness of your gray rhinos, including those gray rhinos disguised as black swans.

Well, that wraps up Chapter 2, "Ten Families of Risk." Next stop is a checklist for success that I've been working on for forty-plus years. This platform allows employees to consistently and accurately perform their activities through five concurrent and interrelated themes that meld risk, systems, service, accountability, and integrity. It provides a reliable process for analyzing tasks and ensures your employees do the right thing, the right way every time. I look forward to examining this concept shortly. Until then, work safely.

Chapter Takeaway on Media Relations/Crisis Communication

Please refer to the media relations/crisis communication material in the Addendum section and ask yourself how you would address the following scenario:

It's 10:30 am, and you receive report of a toxic cloud rising from one of your well sites. Treating wells to keep them clean as well as free of bacteria and viruses is essential. Your organization treats such wells with two separate chemicals. It's a two-stage process, and your employees erroneously reversed the order. The result was a chemical reaction of raw gaseous chlorine. The winds are strong, and the well is adjacent to a grammar school and nursing home. Immediately, you implement your emergency response plan, but it's not detailed or actively practiced. Phone numbers of first responder organizations are limited to the local fire station and 911. There aren't phone numbers for the grammar school or nursing home, and you don't know the principal, medical director, or any of their respective personnel. The fact that there's no contact for your county office of emergency services is equally concerning.

The cloud is small, but the fire department orders an evacuation of the school and nursing home. Parents are at work and worried about their children, as are relatives of the nursing home residents. The community is in a panic even though the event should be over soon without injury. During these tense times, you know your organization will be questioned for its inept response and cleaning practices. Last year, you received a warning from OSHA because three of your employees became dizzy in a similar incident involving raw gaseous chlorine fumes. Fortunately, there was no gas cloud during that incident. This practice is industry standard, but should it be used at well sites within residential areas? You know the answer is no, and you mentally agreed to replace this practice with something safer after the OSHA warning but have yet to act on it.

The phone calls from reporters are inundating the switchboard. You have a good idea of what happened, but how do you explain why it happened? Will the community lose confidence in your organization? What about the board's confidence in you? What if copies of your emergency response plan are requested? How can you defend a plan that has so many holes and was so poorly executed? Do you advise it was on your list to fix, as was outreach to various community stakeholders?

A press conference is necessary, but it's foreign to you. There's no alternative, and the timing is urgent. You must schedule it for this afternoon. First responders are fixing the situation, but how will you address the inevitable reporter questions for the evening news and morning papers? What will you do when the press sets their sights on you and deluge your headquarters this afternoon? The optics are bad, but a cover-up or a drip of facts would be worse. How much information do you share, and how do you prevent being cast as the incompetent character or antagonist in the media's narrative? Do you take questions, read from a script, or offer an extemporaneous explanation of the situation?

3 Five Concurrent Themes for Success

Consistency from Checklists

STRATEGIC HINT FOR YOUR CONSIDERATION

Checklists allow you to distill complex activities (tasks, incidents, and events) into simple, but not simplistic, steps and reminders that promote employee uniformity, consistency, and accuracy throughout one's enterprise.

DOI: 10.1201/9781003229087-3

SUMMARY

Checklists allow employees to distill complex activities into simple steps and reminders. Its core purpose is to prevent tragedies by promoting uniformity and accuracy of activities through coherent and sequential decision making protocol.

Five Concurrent Themes for Success comprise risk, systems, service, accountability, and integrity. This useful checklist allows employees to methodically analyze their discretionary time activities by answering five interrelated questions that apply to all activities performed by your employees:

(1) What's the risk involved in the activity, and how can I best manage it? (2) What's our system, and how can I best implement it? (3) Is there a WOW opportunity? (4) Who's accountable? (5) What are the integrity issues? Once implemented, this efficient checklist can be adeptly completed for all activities.

Gordon Graham here, and welcome back to our journey toward *real* risk management. Chapter 1 described my philosophy as the pursuit by high-reliability organizations to do the right thing, the right way every time through self-reinforcing systems and cultures, with discipline serving as the adhesion. We also expounded on the difference between proximate cause and problems lying in wait as well as the importance of building viable control measures (systems) from root causational factors so future tragedies aren't repeated.

For Chapter 2, we introduced the ten families classification system as a means to properly recognize and prioritize the thousands of risks facing your organization. This system utilizes the investigative tools of risk assessments and risk/frequency matrixes to ensure your catalog encompasses past tragedies, near misses, and industry trends. Its effectiveness necessitates exhaustive and data-centric retrospection, devoid of bias and replete with diversity, to certify that your catalog accurately encompasses all your risks (precept one) so that control measures (systems) are resilient and built around the real or root causes of tragedies.

High-reliability organizations, which espouse *real* risk management, convert their root causational factors into pillars of success. They offer their good people coherent, workable policies (systems) that describe how to act in given situations, meaningful training (initial and ongoing) to perform their activities, effective supervision to promote accountability, and consistent discipline when policies (systems) are violated. Executives, managers, and supervisors are responsible for accentuating these five pillars of people, policy, training, supervision, and discipline. Doing so will augment customer service, minimize civil liability, promote personnel safety, and expand community goodwill.

In this chapter, "Five Concurrent Themes for Success," we'll introduce an overlapping and self-reinforcing checklist that I've been working on for forty-plus years. Checklists allow you to distill complex activities (tasks, incidents, and events) into simple, but not simplistic, steps and reminders. Their core purpose is to prevent tragedies by promoting employee uniformity and accuracy in organizational

Five Concurrent Themes for Success
➲ What's the risk involved in this activity, and how can I best manage it?
➲ What's our organizational system (policy), and how can I best assure its implementation?
➲ Is there a customer service component here, and, if so, how can I maximize it with a WOW factor?
➲ Who's accountable for what on this activity?
➲ What are the issues of integrity involved in this specific activity?

FIGURE 3.1 Five concurrent themes for success. Source: Graham Research Consultants, LLC.

activities through coherent and sequential decision making rules. Prudent checklists and proper decision making are essential components of *real* risk management, as impulsive and injudicious decisions lead to dramatic and permanent consequences (tragedies). Both components necessitate initial and ongoing training to ensure that employees understand how to perform their activities but also why their activities are performed.

The structure for my methodical checklist, which is illustrated in Figure 3.1, comprises five separate and distinct themes that, when put together, allow employees to analyze activities using risk, systems, service, accountability, and integrity. Each activity is analyzed (if there's discretionary time) with the following questions: What's the risk involved in this activity, and how can I best manage that risk? What's our system (i.e., control measures, policies, procedures, processes, rules, checklists, and protocols), and how can I best assure its implementation? Is there a customer service component here, and, if so, how can I include a WOW factor? Who's accountable for what on the activity? And, what are the issues of integrity involved in this specific activity?

Answering these questions will ensure consistent outcomes in organizational quality, responsiveness, and conformance. Checklists provide the discipline adhesion through which employees understand, question, embrace, and enforce organizational systems and their underlying rationale. Discipline cultivates accountability, which is a cure for mediocrity and safeguards sound and ethical decision making. Its derivatives are employee fitness and customer loyalty, which subsequently prevent tragedies. On my recommended reading list is a great book called *Checklist Manifesto* by Atul Gawande. In his introduction, Gawande writes: "The volume and complexity of what we know has exceeded our individual ability to deliver its benefits correctly, safely or reliably. Knowledge has both saved us and burdened us." Gawande's ideas complement the inclination by innovative water/wastewater executives to establish guidelines for the proper exercise of discretion (sound decision making), rather than dismiss its careful application with excessive rulebooks.

STRATEGIC HINT FOR YOUR CONSIDERATION

The volume and complexity of what we know has exceeded our individual ability to deliver its benefits correctly, safely or reliably.

—Atul Gawande (Checklist Manifesto)

There's an inherent value to checklists. In the world of aviation, checklists are a big deal. You show me an aviator who doesn't take checklists seriously, and I'm showing her name on a headstone. Doctors should also be advocates of checklists. In Chapter 1, I talked about medical malpractice and how a minimum of two-hundred-fifty-thousand people are dying annually from such malfeasance. Several thousand patients are killed every year because doctors leave things in them after surgery. That could all be obviated with a checklist. "Doctor, I'm giving you a chainsaw. Before you sew this person up, I'll need the chainsaw back so I can check it off." Obviously, I'm being facetious, but if you place twenty sponges in a patient to control bleeding during surgery, how many do you need to get out? Twenty. One sponge, even in a hospital setting, can be a death sentence. Checklists prevent tragedies, and they improve organizational operations by providing guidelines to do the right thing, the right way, every time. Let's review each of the rules in detail.

THEME ONE

What's the Risk Involved in This Activity, and How Can I Best Manage It?

Step one, risk management, is the cornerstone of the five concurrent themes for success. With that in mind, let's have a brief refresher on *real* risk management. Webster defines risk as the possibility of meeting danger or suffering a harm or loss, or exposure to harm or loss. As a follow then, risk management is the process of looking into the future (short or long term) and asking what can go wrong and then doing something to prevent it from going wrong. Risk is part of life. There's not one of you who hasn't heard the term risk management. Unfortunately, the phrase is grossly misused by people who are misinformed on its depth and scope. The field of risk management goes beyond safety stuff because every identifiable risk is a manageable risk.

The first tranche of questions to ask are the following: Can I eliminate the risk? Why are we doing this activity? Maybe we shouldn't be doing this activity? Can I transfer the risk like the human resources issues we discussed earlier? Risk elimination and risk transfer are forms of risk management. Whatever you're doing and whatever you're thinking about doing must be encapsulated into this core question: What's the risk, and how can I best manage it? Once recognized, you manage the risk by prioritizing it in terms of potential frequency, severity, and available time to think. At that point, you have to mobilize (act) to do something about the recognized and prioritized risk. Mobilization is the systems component, which is the second of my five concurrent themes for success and the underpinnings of *real* risk management. High-reliability organizations do the right thing, the right way every time through a deliberative process of RPM (recognition, prioritization, and mobilization) and an intrinsic culture of employee discipline and continuous improvement. Their propulsion has been and always will be derived from employees. Specifically, employee understanding, adherence, questioning, and enforcement of organizational systems as well as their unequivocal support for the rationale behind these systems.

STRATEGIC HINT FOR YOUR CONSIDERATION

A system is an organized or established procedure or an accumulation of processes as well as a structured approach to doing things. Process and procedure include a series of steps followed in a regular definite order.

THEME TWO

What's Our Organizational System (i.e., Control Measures, Policies, Procedures, Processes, Rules, Checklists, and Protocols), and How Can I Best Assure Its Implementation?

How do you manage risk? You manage risk through the second of five steps on the checklist: systems. Time for a refresher. Webster defines a system as an organized or established procedure or an accumulation of processes as well as a structured approach to doing things. When you check under process and procedure, you'll find a particular way of accomplishing something and a series of steps followed in a regular definite order. Please don't subscribe to the thinking that tragedies can't be prevented. While water/wastewater is a high-risk profession, it isn't the riskiest profession. Timber, aviation, trucking, chemical manufacturers, and construction all have higher adjusted loss rates than your profession. Let's review some of the leaders in those professions. Timber operations are risky, but Boise Cascade is underrepresented in tragedies. Aviation is dangerous, but Southwest Airlines has a phenomenal safety record. Trucking is perilous, but UPS (United Parcel Service) has a low accident ratio in long-haul operations. Chemical plant operations are unsafe, but DuPont is a vanguard in their profession. Construction work is precarious, but Intel has an outstanding lost time injury rate on its projects. Each of these organizations is underrepresented in injuries (tragedies) because they understand the value of *real* risk management.

STRATEGIC HINT FOR YOUR CONSIDERATION

Executives are responsible for building good systems, keeping them current, and performing audits, whereas managers are answerable for implementing systems, and supervisors are accountable for enforcing systems.

Well-designed systems, kept up to date, and fully implemented will never let your organization down. That assertion is also true in water/wastewater operations. High-reliability organizations build and continuously improve their resilient systems and keep them up to date and fully implemented. Please recall Chapter 1. Before I enrolled in law school, I did my graduate work at the Institute of Safety and Systems Management

(ISSM). That curriculum imparted the value and importance of systems. ISSM inculcated the necessity to evaluate systems so they conform to the design, update, implement rule (DUI): properly designed, up to date, and fully implemented. You show me an organization that makes a commitment to systems, and I'll show you an organization that prevents tragedies. On the other hand, you show me an organization that ignores the value of systems, and I'll show you an organization either in trouble or en route to trouble. Executives are responsible for building good systems, keeping them up to date, and performing audits to warrant their full implementation. Managers are answerable for implementing systems, and supervisors are accountable for enforcing systems. I've never done your job in water/wastewater operations, but I've studied your tragedies, and resilient systems are your linchpin. High-reliability organizations are fixated on continuously improving their systems, and that mindset prevents tragedies. You must understand the value of the second of these five concurrent themes for success.

I mentioned earlier that trucking is filled with risk. Some data for you: Eighty-seven firefighters, one-hundred-thirty-three cops, and five-hundred truck drivers died in the United States in 2017. Of the five-hundred truck driver deaths, how many worked for United Parcel Service (UPS)? None. Their drivers aren't dying because UPS made a commitment to systems. I've seen their big rig systems operation. It's impressive and follows the DUI rule: properly designed, up to date, and fully implemented. UPS has a similar system for their feeder operations. DHL also has feeders and so does FedEx. They all know where their packages arrive and where they're sorted. But comparing the lost time injury rates for FedEx, DHL, and UPS, why is UPS so much better? They've made a commitment to systems.

The next time the UPS package car pulls up in front of your building (that's what UPS calls these types of vehicles because it delivers packages), please go to the window and bring a stopwatch. Watch that UPS driver step off the package car. The moment she hits the handbrake, start your stopwatch. She has fourteen-and-half seconds to do eight things. Did you know that? She has fourteen-and-half seconds to set the parking brake, remove seat belt with right hand, remove keys from ignition, rotate keys, open secure cargo door, find package for location, re-secure cargo door, and exit vehicle. Her foot has to hit the ground in fourteen-and-one-half seconds "Gordon, why the half second?" When you have scores of stops a day and several thousand trucks, that half second is a big deal. When she exits the vehicle, watch her feet. If she's carrying the load in her right hand, then she'll be leading off the truck with her left foot. If she's carrying the load in her left hand, then she'll be leading off the truck with her right foot. There are human factors behind this process. It's safer and easier to lead with the opposite foot when carrying a load.

STRATEGIC HINT FOR YOUR CONSIDERATION

High-reliability organizations are fixated on continuously improving their systems, and that mindset prevents tragedies and promotes organizational mission, purpose, and success.

Watch her lay down your package. You're not going to see any back action. It's leg action. She's briskly returning to the truck now. Watch her left foot. I've enjoyed many free lunches from lawyers in downtown Los Angeles with this one piece of data. For almost forty years, I've been going to lunch with lawyers. Hey, Tommy. "Gordy, what's up?" Take a look at the UPS driver returning to her truck. I bet you lunch her left foot hits the bottom step first. And up goes the left foot. "Gordon, you lucked out again." Forty years in a row. I'm on a streak. Left-right-left and sit-start. When you're entering/exiting a truck scores of times a day, you need a process on how to do it. Watch how quickly she buckles her seat belt. Incidentally, it's a big deal if a UPS driver is caught not wearing her seat belt. Watch how she performs a circle check around the vehicle to make sure there are no people or obstacles before she backs up. "Gordon, are you trying to tell me something?" Oh, how I'd like to study your accumulated data. "Gordon, in our organization, vehicles were involved in one traffic accident for every one-hundred-and-thirty-seven-thousand miles we drive forward, but we have one traffic collision for every fifty-thousand feet we back up. We keep backing into things. Why?" It's all about systems, but they must be properly designed, kept up to date, and fully implemented.

I'm concerned about the design of your systems and if they're up to date, but I'm going to suppose right now you have well-designed systems that are kept up to date. That, however, isn't good enough. The California Highway Patrol (CHP) has a brilliant pursuit policy. How do I know? I wrote the policy. It's properly designed and up to date. Then why does the CHP keep getting into trouble with their pursuits? Because people don't follow rules. I know organizations with brilliant sexual harassment policies that are properly designed and kept up to date. Then why have they paid so many millions of dollars in harassment claims? Because people don't follow systems, and why don't people follow systems? Take a look at Figure 3.2. Three words: arrogance, ignorance, and complacency.

STRATEGIC HINT FOR YOUR CONSIDERATION

What's arrogance? Rules don't apply to me. What's ignorance? I had no idea we had a rule on that. What's complacency? We've been doing it this way for years and haven't had any problems.

Three Reasons Personnel Don't Follow Systems

- Arrogance
  (Rules don't apply to me)
- Ignorance
  (I had no idea we had a rule on that)
- Complacency
  (We've been doing it this way for years and haven't had any problems)

FIGURE 3.2 Three reasons personnel don't follow systems. Source: Graham Research Consultants, LLC.

What's arrogance? Rules don't apply to me. What's ignorance? I had no idea we had a rule on that. What's complacency? We've been doing it this way for years and haven't had any problems. You can be arrogant, ignorant, and complacent in many professions, but it's a problem lying in wait in water/wastewater operations. So with that in mind, does Bud work for you? "Gordon, who's Bud?" Bud is the employee who's absolutely convinced the rules don't apply to him. Bud is the personification of arrogance, ignorance, and complacency. And there are Budettes who also think the rules don't apply to her. Bud and Budette are problems lying in wait. "Well Gordon, I find that offensive. There are thousands of names. Why did you single out Bud?" Czar 52. Research Czar 52. It's the story of a preventable B52 crash in Washington State. If you go to YouTube, you'll get lots of different hits. Click on any one of them. They're all the same. The first thing you'll see on your screen is a date: June 24, 1994. The second thing you'll see is a location: Fairchild Air Force Base, Washington. The third thing you'll see is this beautiful B52 making lazy circles around Fairchild Air Force Base in preparation for an air show the next day. Depending on which video clip you watch, that beautiful B52 slides out of the sky, into the ground, blows up, and kills everybody on board. Who's flying the plane? Lieutenant Colonel Bud Holland.

That's a tragedy. The Air Force investigated and part of their protocol is to interview every pilot on the base. Every pilot said the same thing: "It wasn't a matter of if, it was a matter of when, where, and how many people were going to die." This guy had a long history of reckless behavior. He was a rogue pilot. On my recommended reading list is a great book by Tony Kern, one of the smartest people I know. Tony was a B1 pilot and picked up his doctorate on the side. His great book, *Darker Shades of Blue*, studies tragedies in the Air Force. Czar 52 is a focus throughout his book. Bud Holland had a long history of reckless behavior. Everybody knew he was a rogue pilot who acted as if the rules didn't apply to him. Go back to what he did in 1990 during a bombing exercise over Guam as well as various airshows in 1991, 1992, and 1993. Everybody knew he was a rogue pilot, a reckless aviator. Nobody did anything about it until after the tragedy.

STRATEGIC HINT FOR YOUR CONSIDERATION

If you ignore Bud, other people will notice it; and the deviation from the norm will become the norm. "Bud did it. Why can't I?" You must address the Buds in the workplace. They're problems lying in wait.

I put a new book on my recommended reading list because it delved deeper into Czar 52, *Warnings Unheeded* by Andy Brown. Five days prior to the B52 crash, an airman went into the base hospital at Fairchild with an automatic rifle and killed several people. What did we learn after the fact? Air Force psychiatrists were telling base management that this airman was unstable, a problem lying in wait. There was advanced warning he was going to kill people, and the warning was ignored. So with that in mind, I have a question for the executives, managers, and supervisors: Does Bud work for you? "Well, Gordon, how can I possibly identify Bud?" Let's go back to when you were a line employee six months into the job. How did you look at your coworkers back then? Did you know the Buds in your organization? Did you know who were arrogant, ignorant, and complacent? Yes, you did. Now you're in charge? Are you uninformed all of a sudden? No. You know the Buds in your organization. It's a matter of will. If you take Bud on, you get paid X. If you ignore Bud, you get paid X. So why take him on? Because that's your job as an executive, manager, or supervisor. You must identify the Buds and address them proactively. If you ignore Bud, other people will notice; and the deviation from the norm will become the norm. "Bud did it. Why can't I?" You must address the Buds and Budettes in the workplace. They're problems lying in wait.

THEME THREE

Is There a Customer Service Component Here, and, if so, How Can I Maximize It on This Activity?

Theme one: How do you manage risk? Theme two: You manage risk through systems. The third of five concurrent themes for success is the importance of customer service. Theme three is a critical issue today, and many water-related entities don't understand the ramifications of not taking it seriously. I'll give you a blanket statement: Americans are fed up with the lack of customer service. "Gordon, you make these blanket statements. Do you have data to verify its veracity?" I do. Have you stopped doing business with a company because they angered you just once? Be honest. My conjecture is yes. Let me ask a second question: Did you keep your anger secret? My guess is no, and I'll presuppose you were so angry that you shared your dissatisfaction with other people and encouraged them to do the same. Successful organizations have figured that out. If you make enough people angry, you're going to lose market share. On the other hand, if you want to increase market share, you must make customers happy. High-reliability organizations focus on customer service. Water-related entities must adopt this type of thinking.

Research customer service and you will find articles on Nordstrom's, the department store. Nordstrom's has a peerless reputation for customer service. "Gordon, they also make excellent clothes." No, they buy clothes from clothing manufacturers and sew Nordstrom labels on them. Men's suits at Nordstrom's are expensive. You can buy a similar suit elsewhere for much less. How does Nordstrom stay in business? Service. Go down to a department store tonight and buy a suit. "So why are you buying a suit, my friend?" My sister is getting married on Saturday, and my wife suggested I buy a new suit. "That could be problematic because suits purchased today have an alteration turnaround of four weeks from purchase." But I need it by Saturday. That exchange doesn't do the customer any good.

Go to Nordstrom tonight, and buy the suit. They'll alter it while you wait, and if you can't wait, they'll deliver it to your house and follow-up to make sure the suit fits perfectly. If the suit gets torn at the wedding, Nordstrom will take it back and replace

Customer Service Ingredients

Technical Competence + Knowing Your Job + Dignity & Respect + WOW

= Loyal Customers

FIGURE 3.3 Customer service ingredients. Source: Graham Research Consultants, LLC.

it. No questions asked. Nordstrom has mastered the value of customer service. Starbucks has also figured it out. People are paying five dollars for a cup of hot water because Starbucks knows the value of customer service. Lexus has figured it out, and they're simply Toyota with a high level of customer service and a burnished brand. Service is absolutely essential, and there are four core components to creating loyal customers. Satisfied customers aren't enough. You need loyal customers who'll continue to do business with your organization and rave about your service. Figure 3.3 provides the four core component formula for achieving customer loyalty.

Too many organizations stress the importance of customer satisfaction. What does that mean? Satisfied customers may come back and do business with you again. That's not enough. I want loyal customers because loyal customers will come back, and they'll bring others with them. So how do you create a loyal customer base? Executives must clearly understand their customer base, and they must secure the necessary resources to address their needs. That's where it typically ends for most water-related entities. You must do a lot more to achieve customer loyalty, and it starts from the bottom up. Every contact from every employee is an opportunity to create loyal customers. My customer loyalty program has four principle components: (1 and 2) doing the right thing, the right way every time (technical competence + knowing job); (3) treating people right (dignity and respect); and (4) adding in the WOW factor. Let's take these one at a time.

STRATEGIC HINT FOR YOUR CONSIDERATION

Doing the right thing, the right way every time, treating people right, and adding in a WOW factor create customer loyalty and that requires a clear understanding of customer needs as well as the necessary resources to address those needs.

Doing the right thing, the right way every time: It's imperative the activities your people perform are done the right way every time. Most of what your employees do, they do right. Notwithstanding, you must be cognizant about the dangers of complacency, fatigue, distraction, hubris, and risk homeostasis as well as the importance of regular training for high-risk, low-frequency events (core critical tasks and critical tasks). *Treating everyone involved with the highest level of dignity and respect possible under the circumstances*: Technical competence and knowing your job is no longer good enough. It must be complemented with dignity and respect to ensure your employees treat people properly in their contacts. Unfortunately, there's a lack of dignity and respect in the United States. The simple things we took for granted in decades past are now virtually non-existent. Holding doors, saying please and thank you, smiling, and properly addressing people are disappearing. Where did the employee learn that behavior? You didn't teach your people to be rude and uncaring. Like anything else, it's learned long before they came into your profession. We must work to improve the way people are treated, and that requires revisiting rules of decency such as treating people the way you'd like to be treated, smiling, refraining from negativity, communicating the purpose of your particular activity, apologizing when you make a mistake, listening, and following through on your promises.

STRATEGIC HINT FOR YOUR CONSIDERATION

If your response to customer service is it doesn't apply to water-related entities or isn't applicable to my position, then you're in the wrong profession, as every organization and every position involves customer service.

Add in the WOW factor whenever you can: So what's this WOW factor? It's simple and was developed by the late Chief Brunacini of Phoenix Fire Department. Every external contact should conclude with the customer or member of the public saying WOW. Here are some areas where you can create WOW: *Initial greetings with people you contact*: There should be courtesy, dignity, and respect at the start to every successful contact. "How are you doing today?" is a nice start to an activity. Saying please and thank you and answering questions are also important. *Telephone issues*: All calls should be answered in a timely manner, and your calls should be coherent as well as polite. You should listen carefully and direct calls to appropriate parties, when necessary. It's important to regularly check your voicemail and promptly respond to calls made to you. If you're going to be absent for a period of time, put that on your message and give the caller another option. *Vehicle operations*: A lot of public opinion is developed by the way you drive your organization-decaled vehicles. Drive within the law and policy, keep your equipment clean, use seatbelts and turn signals, acknowledge courteous driving behaviors, and park in designated spaces. *Response to calls and other events*: Timeliness is essential. If you've made an appointment, be there on time. If something unexpected comes up, inform the party you'll be late. Nothing vexes people more than being inconvenienced. *Following up after the initial contact*: Check-in after the contact to see if there's anything else you can do to assist the person you contacted. You'll be amazed at what you hear from the people you contact. *Dealing with bad customers*: Some contacts are with problem people who'll never be pleased, but it's still important to treat them well. They might be a juror in a lawsuit or a project advocate someday.

When I address the topic of WOW with water/wastewater professionals, I'm often greeted with the following: "Well, this degree of engagement may apply to some jobs, but as a (insert job description), it isn't applicable to me." If you are of that belief, hopefully I can change your mind. Allow me to use a public safety example to accentuate the point. I dread receiving a jury summons and not because I don't want to serve on a jury. The reason is a lack of service to people who serve as jurors. Let's start with the initial phone call to the jury commissioner's office for a change of date. I've personally experienced long hold times, disconnections, rude call handlers, and many other problems. Now let's fast forward to the date of your jury summons beginning with remote parking for the jury pool. We then enter the courthouse and frequently interact with disheveled and rude deputies who are counting down their retirement. Some of the potential jurors have never been through security screening and don't understand the process. Discourteous behavior isn't going to make them understand the process if they haven't done it before. What's ironic is the uniform worn by an impolite security deputy may well be the same uniform worn by a street deputy who's testifying in a criminal case. That baggage may sway juror opinion of the street deputy's testimony. And I could detail the courtroom proper, from insolent prosecutors to condescending judges who behave with impunity. I'd like to think my experiences are isolated, but they're consistent with other people's experiences when I talk to them.

STRATEGIC HINT FOR YOUR CONSIDERATION

Enhancing customer service involves identifying specific behavior that made you angry and not participating in those behaviors. Simultaneously, it involves doing the things that made you say WOW.

For your average American, they've been paying taxes their entire life and haven't had experience with the criminal justice system. They're trying to do the right thing by accepting the jury summons and not pulling an excuse. These are good Americans who are agreeing to participate in the process, and they're not treated properly. Please recall that this interaction may be their only experience with the criminal justice system. They won't keep their anger private. They'll tell other people who'll tell other people. And that's how you earn a government approval rating of nineteen percent. Let's change the scenario and apply it to water/wastewater operations. How are your customer interactions as it relates to shutoffs, repairs, breaks, permits, billings, governance, vehicle operations, and community relations? These touch points impact your goodwill and invariably influence your future budgets as well as bond measures and outcomes downstream. The public is fed up with unacceptable customer service, and they aren't being silent about it. They vote, either with their feet by abandoning organizations that don't provide service or in a polling booth (or jury) when water-related entities (and governmental organizations) don't take it seriously.

STRATEGIC HINT FOR YOUR CONSIDERATION

Your imagination (within the confines of your policy manual) is the only limit for excellent customer service. Be nice and use your head to create WOW with as many people as you can.

How many of you have had a negative experience with an organization that made you so angry you vowed never to do business with that organization again? I'll ask a second question. How many of you have had a positive experience with an organization that made you say, WOW? Take a moment and list the three most negative contacts you've had with any organization. Identify the specific behavior that made you angry, and list that behavior. As simple as it seems, don't participate in those behaviors. Simultaneously, do the things that made you say WOW. Take a look at the lists from your fellow employees. Talk about the things that create loyal customers. If you don't feel that this level of exceptionalism is within the purview of your job, then perhaps you've chosen the wrong line of work. If your paycheck is ultimately charged back to ratepayers, then you work for the community and are a public servant. It's easy to make every contact an opportunity to create a loyal customer. Your imagination (within the confines of your policy manual) is the only limit. Be nice and use your head to create WOW with as many people as you can. They're all jurors, voters, and members of our great public. You have a key role here and taking customer service seriously will have major benefits in the future.

THEME FOUR

Who's Accountable for What on This Specific Activity?

Theme one of my five concurrent themes for success is managing risk, and you do that through theme two, properly designed systems that are kept up to date and fully implemented. Theme three is the importance of customer service, which comprises technical competence, knowing your job, dignity and respect, and adding some WOW. The fourth theme is the dying word known as accountability. I say dying word because wherever I go around this great country and talk to people in depth about issues, they always mention the dearth of accountability throughout our society. Statements such as "That's not my job" and "She doesn't work for me" and "We've never done it that way before" are tiresome. Customers are exasperated with employees who are unwilling to accept the fact that they need to be accountable.

If you want accountability, it starts when everybody in your organization understands their role with respect to systems. For the executives, your accountability is building good systems and keeping them up to date. Your accountability continues and includes an expansive and robust audit process to make sure what you say you're doing is in fact being done. The key word is robust. Too many organizations have perfunctory audits. There are all sorts of documents saying everything is wonderful when in reality there are significant problems lying in wait. Your audits and inspections should encompass all aspects of your organization. These tools form an integral part of an executive's role and accountability with respect to systems. You can't assume all is going well. There must be control measures (systems) in place to assure things are being done right. That's not micromanagement. It's called doing your job. We need a feedback loop in every organization, and that's the role of audits. If you don't have the audits (formal and informal) in place, then you won't know about problems until they become consequences, and then you're in the domain of lawyers. At that point, it's too late for action. All you can do is address the consequences.

STRATEGIC HINT FOR YOUR CONSIDERATION

Show me a tragedy in a water-related entity and one of the real problems lying in wait will be a supervisor not behaving like a supervisor, or a supervisor who tried to behave like a supervisor but wasn't supported by management.

Formal audits are imperative, but don't forget the value of informal audits through management by walking around (MBWA) and supervising by walking around (SBWA). MBWA and SBWA lead to proper conduct. Take a look at what your people are doing. Walk around, and you'll quickly learn your people are doing most things right. The top complaint I get from line-personnel is the following: "The only time I ever hear from my supervisor is when something is wrong." Please work to change that perception. On those rare occasions, when people aren't following policy, it must get addressed through the discipline process. Executive accountability continues after you complete your audits, inspections, and complaint investigation. It involves promoting women and men into the ranks of management and supervision who have the fortitude to be managers and supervisors. I've dedicated an entire upcoming chapter on the importance of supervision and the key role of the supervisor.

STRATEGIC HINT FOR YOUR CONSIDERATION

You must promote good women and men who have the fortitude to be supervisors, and then you must support these supervisors when they make the tough calls.

Show me a tragedy in a water-related entity and I'll show you a proximate cause of X. All too often, the real problem lying in wait is a supervisor not behaving like a supervisor. Or in the alternative, a supervisor who tried to behave like a supervisor but received no support from management. An upcoming chapter will focus on job-based harassment, and a substantial area of concern is a subset called hostile work environment. Executives, managers, and supervisors are all familiar with allegations of hostile work environment. What's the definition of a hostile work environment? Severe, pervasive, pattern, and practice of misconduct directed at protected class status. Here's a thought for you: How do you have a severe, pervasive, pattern, and practice of misconduct occurring in the workplace if supervisors are behaving like supervisors? I say the same thing about California Highway Patrol (CHP) pursuits. Many a times the policies aren't being taken seriously. I wrote the policy. I know what it says. There are too many units chasing registration violations. There are too many units on motorcycles chasing violations that aren't allowed by the policy. "Sergeant, why didn't you call it off?" "I don't like to second-guess my people when I'm not on scene." That's an excuse for lacking the courage to enforce the rules.

Supervisors must behave like supervisors, and you must promote good women and men who have the fortitude to be supervisors. You then have to support these supervisors when they make the tough calls. Executive accountability continues with having a discipline process in place to address the Buds (the arrogant, ignorant, and complacent). You must address these people proactively because Bud can't survive in your high-risk profession. More on discipline in a later chapter. That's executive accountability. For managers, their accountability is implementation of systems. The primary mission of a manager in any profession is systems implementation. For supervisors, they arguably have the hardest role. Their accountability is enforcement of organizational policy or making sure people follow rules. I'll have a chapter on the key role of the supervisor, and I'll give you some rules for improving supervisory performance. The primary mission of a supervisor is enforcement of organizational policy and making sure your people know the policies and follow the policies (systems). When it comes to line-personnel, their accountability is knowing the policies (systems) that apply to their job and following the policies. Not some of the time, not most of the time, but all of the time. Know all the policies, and that includes the seatbelt policy, the self-contained breathing apparatus (SCBA) policy, the personal protective equipment (PPE) policy, the backing policy, the workplace violence policy, the harassment policy, and all the others. Line-personnel must follow the policies of your organization.

STRATEGIC HINT FOR YOUR CONSIDERATION

A lack of integrity is a problem lying in wait because its removal causes mediocrity, and mediocrity is a cancer that overpowers accountability and quickly destroys organizations.

One final thought on accountability. I talk to all sorts of groups, and I always mention a cop story to every high-risk profession including water-related entities. Los Angeles Police Department (LAPD) experienced reputational injury and loss of public trust in the late 1990s with the Rampart Scandal. Research the LAPD Rampart

Scandal. One-hundred-fifty-million dollars of civil liability, with thirty cops incarcerated or dismissed. There was major organizational embarrassment. After the tragedy, lawyers were assembled and a seven-hundred-eighty-two-page final report was compiled. In a typical lawyer fashion, they focused on proximate cause. "Oh, this is wrong with LAPD. This is wrong with LAPD. And this is wrong with LAPD." One smart guy said: "You're looking at it the wrong way. Let's look for the real or root problem with LAPD." And he summed it up in four words: "Mediocrity has replaced accountability."

His brilliant quote ended up on page one of the final report:

> Commander Ross Swope, Metropolitan Police Department Washington, D.C. (DCPD) went on to explain that mediocrity stems from a failure to hold people responsible and accountable. It comes from a lack of commitment, laziness, and excessive tolerance. He said mediocrity is the greatest challenge we face.

The following passage is taken verbatim from the LAPD report published in March 2000. Every employee in every water-related entity must understand these words and what they mean to your operation:

> Captain Swope went on to explain that mediocrity stems from the failure to hold officers responsible and accountable. It comes from a lack of commitment, laziness, excessive tolerance and the use of kid gloves. He felt that dealing with mediocrity is perhaps the greatest contemporary challenge to American police departments. When asked to explain how mediocrity is dangerous, Captain Swope drew an analogy of the bell curve. At the high end of the bell curve are those officers who practice all the core values: prudence, truth, courage, justice, honesty and responsibility. At the other end, are the officers with few of those values. In the large middle are those officers who have some or most of the core values. The extent of moral influence in a police department depends on the extent to which the upper and lower portions influence those in the middle. The men and women who control that influence are sergeants, lieutenants and captains. The irony is that everyone within a workplace knows full well which of the three categories their co-workers fall into. When officers in the middle see that officers at the bottom end are not dealt with, they sometimes begin to imitate their behavior. Similarly, when those at the top end are recognized and rewarded, they become the workplace standard. The principal, though not exclusive, agents in encouraging top end or bottom end behaviors are supervisors and middle managers. It is our sergeants, lieutenants and captains who have the daily and ongoing responsibility to ensure that appropriate workplace standards are maintained. However, that observation in no way relieves upper managers from their responsibility to ensure that proper standards are being maintained in their subordinate commands by providing appropriate guidance, exerting their oversight responsibility and honestly evaluating the effectiveness of the commands for which they are ultimately responsible.

The essence of Commander Swope's statement is personnel follows culture, and culture begins with your executives and continues through its implementation by your managers. It's your supervisors, however, who ultimately determine the true organizational culture by ensuring consistent enforcement of prescribed systems.

It's those women and men who have daily and ongoing responsibility to assure that appropriate workplace standards are being maintained. The key role of the supervisor is enforcement of organizational policy. On any given day, nobody knows where the executives are, and nobody knows where the department managers are, but everybody knows the location of their supervisor. And riddle me this: Will some employees modify their behavior based on which supervisor is on duty? Or as my grandmother used to say: When the cat's away, the mice will play. Supervisors must enforce policy. Everyone in the organization, up, down, and around the chain of command has accountability. There are different levels of accountability, but everyone is accountable for doing their jobs correctly. When accountability isn't present, you have mediocrity. Mediocrity is a cancer that spreads quickly in organizations, and, if not eliminated, it will overpower accountability and destroy your integrity.

THEME FIVE

What Are the Issues of Integrity Involved in This Specific Activity?

And the fifth of these five concurrent themes for success is integrity. Lose integrity and you won't achieve its manifestation known as ethical behavior. Lose your ethical behavior, and you've lost the public trust. Without the public trust, you have nothing. When you look at stories about public trust in water-related entities, it's not always good. Everything we do has to be at the highest levels of ethics and integrity. There are no more secrets. The public, your customers, demand full transparency. Where did this slippery slope start? It starts with the executives, managers, and supervisors in your organization. If they're counting the days to retirement, then they need to leave now. Customers have the right to expect integrity from your personnel and not just at point of hire but throughout their employment career. And that goes back to the importance of recruitment, backgrounds, probation, and supervision. How much damage could one bad employee in your organization do? Putting up with a lack of integrity is a problem lying in wait.

STRATEGIC HINT FOR YOUR CONSIDERATION

Lose integrity and you won't achieve its manifestation known as ethical behavior. Lose your ethical behavior, and you've lost the public trust. Without the public trust, you have nothing.

STRATEGIC HINT FOR YOUR CONSIDERATION

Completing assigned activities the right way, the first time, while treating all with dignity and respect, should be the goal of every employee.

FINAL THOUGHTS

Water/wastewater operations are complex in nature and full of risks. It takes a good person to be a good water/wastewater professional. Being a good person, however, isn't enough. In order to be ethical and professional, all water/wastewater professionals must receive a coherent, workable policy manual (systems) that provides guidance on how to act in given situations. Additionally, good people must be fully trained to accurately perform every aspect of their rightful activities. Training is an everyday event and must focus on high-risk, low-frequency activities (core critical tasks and critical tasks). Supervisors and managers must similarly be accountable to warrant that all employees are doing the right thing, the right way every time. When they aren't, there must be discipline to address the violation, not because of consequences but because of deviation to policy.

Completing assigned activities the right way, the first time, while treating all with dignity and respect, should be the goal of every employee. Your managers and supervisors have the daily obligation to ensure employees are fulfilling organizational mission and conforming to prescribed systems. Success will augment customer service, minimize civil liability, enhance community goodwill, and protect your personnel. Once fully implemented, employees will be able to adeptly and expeditiously utilize the five concurrent themes for success in their daily activities.

There you have it: my checklist. Theme one: What's the risk involved in what we're doing, and how can I best manage that risk? Theme two: What's our organizational system, and how can I best assure its implementation and enforcement? Theme three: Is there a service opportunity here, and how can I provide a WOW factor? Theme four: Who's accountable for what on the activity? And theme five: What are the issues of integrity involved in this activity? When these five themes become fully inculcated into your organization (risk, systems, service, accountability, and integrity), your employees will do the right thing, the right way every time. When things go right, we stay out of trouble. The anchor for these five concurrent themes is the practice of theme one, *real* risk management, which is predicated on three precepts: (1) There are no new ways to get in trouble; (2) there's always a better way to stay out of trouble; and (3) predictable is preventable.

Our next chapter, "Organizational Risk Management," will encapsulate the imparted lessons from the first three chapters and form the learning objectives for the rest of the book. The essence of *real* risk management is identifying and addressing the root causational factors of your problems lying in waiting and converting them into pillars of success. Good people who act with good policy, who are regularly trained, properly supervised, and in an organization that addresses arrogance, ignorance, and complacency with fair and impartial discipline will prevent tragedies by doing the right thing, the right way every time. That correlates to the self-reinforcing values of culture and systems, which produces consistent employee behavior through conformance with and knowledge of systems. Its derivatives are an underrepresentation in tragedies and overrepresentation in employee fitness and customer loyalty. Next stop: organizational risk management. Until then, work safely.

Chapter Takeaway on Media Relations/Crisis Communication

Please refer to the media relations/crisis communication material in the Addendum section and ask yourself how you would address the following scenario:

You arrive at your headquarters at 7:30 am only to be greeted by angry senior citizens as well as throngs of news and print media. These senior citizens are upset because your organization deactivated water service to several elderly ratepayers. The ratepayers include a military veteran, school teacher, and community nurse. Most are on fixed incomes and struggling to pay the increased rates required to maintain safe and reliable drinking water. The elderly ratepayers are paying the old rate, but you had to reject their payments to avoid any appearance of selective enforcement. The problem became exacerbated when one of the ratepayers went two days without any water. Fortunately, a neighbor sensed a problem and alerted her family. Crisis averted. Thank goodness.

You know your position is technically correct, but the optics are horrible. Why didn't your employees alert you of the situation? You have specific policies to address those who are indigent, but the elderly ratepayers didn't meet that definition. There was also no communication by your employees to defuse the situation and work out a mutually acceptable plan. Your staff followed notification protocol, but there was no instituted checklist for assessing the risk, examining policy, offering customer service, maintaining accountability, and ensuring integrity. This type of checklist was being used by a couple of neighboring organizations, but you dismissed it as being unnecessary. We have good people and good policy. They're properly trained and supervised, and we apply discipline when it's needed. There's no need for a checklist when your people have common sense.

It's clear a properly designed checklist to promote employee consistency and perspective would've prevented this organizational embarrassment. Prudent discretion was needed, and there's no excuse for not seeing the value. All these thoughts are crossing your mind, and you know the community will be rightfully upset. Equally concerning, the media will make you out to be the villain. How do you explain the situation while being compassionate and honest? How do you protect your organization's goodwill as well as your governing body's reputation for being fair minded and not parsimonious? Do you cast blame on your supervisors for mishandling the situation? Or do you own the blunder and accept the consequences?

4 Organizational Risk Management

Applying Theory to Practice

2019 Photo Courtesy of Yorba Linda Water District

STRATEGIC HINT FOR YOUR CONSIDERATION

Training is the overarching pillar that influences the other four pillars while concurrently serving to prevent tragedies through mission achievement, employee fitness, and customer loyalty.

DOI: 10.1201/9781003229087-4

SUMMARY

Organizational risk management is where the theory of real risk management is applied to practice. This theory encompasses the prevention of tragedies through actionable control measures (systems) that transcend proximate cause by addressing the real problems lying in wait.

High-reliability organizations convert their root causational factors, or problems lying in wait, into pillars of success by getting and keeping good people, designing good policy, building solid, realistic, verifiable, and ongoing training, providing adequate supervision, and enacting a fair discipline process.

Chapter emphasis will center on the pillar of training, which underpins and augments the other pillars of success by providing employees a viable framework to do the right thing, the right way every time so that organizational mission and customer loyalty are attained.

Gordon Graham here, and welcome to the seminal chapter in our journey toward *real* risk management. Chapter 1 described the essence of my philosophy, which is the inexorable pursuit by high-reliability organizations to do the right thing, the right way every time through self-reinforcing systems and culture, with discipline serving as the adhesion. Discussion also centered on the distinction between proximate cause and root cause of tragedies. Chapter 2 introduced the importance of the ten families classification system and its investigative tools of risk assessments and risk/frequency matrixes. These tools recognize and prioritize risks to ensure your mobilization (systems) component is aptly built on root causational factors. Chapter 3 showcased the five concurrent themes for success, which offer an actionable checklist for employees to consistently, correctly, and collectively perform their activities while concomitantly achieving organizational mission.

Chapter 4, "Organizational Risk Management," functions as the underpinning of my book and is where the theory of *real* risk management is applied to practice. It's the determinative initiation of an enterprise's ability to manage risk and ascend to a high-reliability organization. Such rite of passage requires purposeful handling of five interrelated and universal subsystems of people, policy, training, supervision, and discipline. We peripherally examined these subsystems and will dedicate this chapter and those that follow to delving deeply into each of the self-reinforcing pillars. Emphasis will center on training, as this overarching pillar influences all other pillars while concurrently serving to prevent tragedies through mission achievement, employee fitness, and customer loyalty. Let's provide a foundational refresher on organizational risk management so we can proceed to a more granularly examination of its subsystems.

Organizational risk management is an enterprise belief that bad things (tragedies) are preventable in water/wastewater operations. This conviction is predicated on the three precepts of *real* risk management: (1) Past tragedies are predictive of future tragedies; (2) there's always new ways to prevent tragedies through system improvement; and (3) identifiable risks are manageable risks. All subsequent theories are yielded from these precepts, including conversion of root causational factors into

Organizational Risk Management
Root Causes = (1) People + (2) Policy + (3) Training + (4) Supervision + (5) Discipline
High-reliability organizations identify and correct their root causational factors (problems lying in wait), which then allows for the conversion of their problem factors into solution factors and the manifestation of five pillars of success.

FIGURE 4.1 Organizational risk management. Source: Institute of Safety & Systems Management (ISSM).

pillars of success. This conversion involves investigating tragedies beyond proximate cause, which is the event immediately preceding the tragedy, by identifying problems lying in wait (root causational factors) that people knew or should've known about but no corrective action was taken.

Once the principle of root causational factors is understood, you can effectively practice the notion of RPM (recognition, prioritization, and mobilization), which is also yielded from these three precepts. We know RPM involves the recognition and prioritization of risks as well as mobilization (action) through properly designed, up-to-date, and fully implemented systems. The five root causational factors of people, policy, training, supervision, and discipline are invariably involved in tragedies. These problem factors can be converted to solution factors through a theory known as analogues or opposite effects. The theory's base involves addressing and correcting your root causational factors so they transform into pillars of success. Such transformation encapsulates the essence of high-reliability organizations. The success of these enterprises is attributed to getting and keeping good people, deriving and maintaining good policy, ensuring adequate and ongoing training on policies, having appropriate supervision of employees to enforce policy conformance, and taking appropriate discipline when there's a deviation from policy. Let's now look at our familiar Figure 4.1.

STRATEGIC HINT FOR YOUR CONSIDERATION

The theory of analogues, or opposite effects, allows problem factors to be converted into solution factors by addressing and correcting root causational factors so they transform into pillars of success.

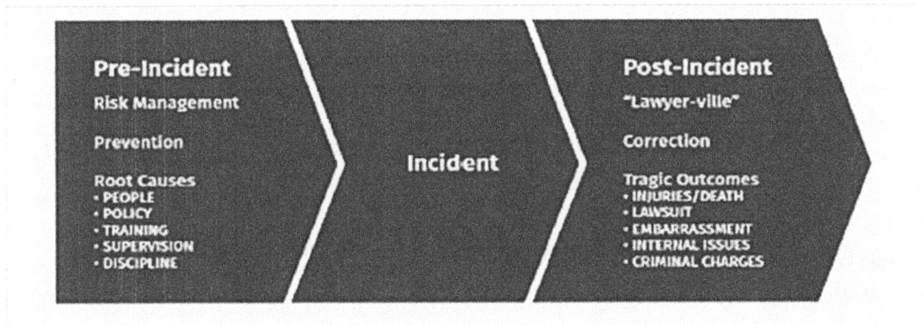

This recognizable graphic visualizes the entire chapter. On the right side is the area I call Lawyerville. Lawyers handle tragedies after they occur. They handle the deaths, injuries, embarrassments, lawsuits, internal investigations, and rare criminal filings. Risk managers study tragedies, and we look for root cause, otherwise known as problems lying in wait. The lawyers will tell you there are thousands of causes because their bias is proximate cause. You ask a real risk manager how organizations migrate into Lawyerville, and she'll respond by saying that there aren't thousands of causes but five root causational factors of tragedies. I've never done your job in water/wastewater operations, but I've studied your tragedies, and most, if not all, link back to the root causational factors of people, policy, training, supervision, and discipline. Let me condense the learning objective of this chapter right now: You give me good people who have good policy, ongoing training, supervisors behaving like supervisors, discipline when rules aren't being followed, and you're going to prevent tragedies by doing the right thing, the right way every time. We're now ready to delve into the details and examine each of the five pillars of success that comprise organizational risk management.

PILLAR ONE

PEOPLE

STRATEGIC HINT FOR YOUR CONSIDERATION

The easiest way to increase the size of your applicant pool is to lower standards, and every time you lower standards, you'll pay for it downstream. An applicant puddle is a problem lying in wait.

Pillar one is people. It's important to point out the obvious: Not everyone is qualified to be a water/wastewater professional. You need a front door to screen out those who are unfit and a back door to remove those who snuck in the front door. The front and back doors are analogies for recruitment, backgrounds, probation, and performance evaluations. I'm going to thoroughly discuss background investigations and performance evaluations in Chapters 5 and 6. So that leaves recruitment and probation. We hire from an applicant pool. If your applicant pool is an applicant puddle, that's a problem lying in wait. I can't speak for your organization, but there's a shortage of qualified workers throughout the United States. It's difficult to find good people, and the easiest way to increase the size of your applicant pool is to lower standards. Every time you lower standards, you'll pay for it downstream. You don't have to lower standards, but you must revisit your recruitment policy. I know recruitment varies by the size and the location of the organization. Those metrics are irrelevant in this discussion. Once again, you have a problem lying in wait if your applicant pool is an applicant puddle.

There's a way to increase the size of your applicant pool without lowering standards. The answer is to make every employee a recruiter. If I were your executive-in-charge, I'd amend the policy manual to require all employees on duty to recruit. This wording has to be clear in intent and carefully reviewed by competent counsel, but it's necessary. Let's think about the numbers. If everybody in your organization made it their goal to find one good woman and one good man in their career, that would address attrition. If everybody in your organization made it their goal to find one good woman and one good man per year, you'd have a wide applicant pool. And if everybody in your organization made it their goal to find one good woman and one good man per month, you'd have a wide, deep, and talented applicant pool from which to select the best-of-the-best. "Gordon, what do you think we need as a water-related entity?" Two things: (1) top-notch people; and (2) top-notch people who look like your community. If you don't have both, you have a problem lying in wait. Recruitment is an obligation for all of us. Encourage your people to actively recruit. They're your best solution to this problem, and make sure your applicant pool reflects your community. To not have proper recruitment is a problem lying in wait.

The takeaway is simple: Everybody must be a recruiter. "Gordon, do you actually practice this stuff?" Yes, I do. I'm the number one recruiter for the California Highway Patrol (CHP). Where do I recruit? I'm on airplanes a couple hundred times a year. You're sitting next to somebody twenty-one to thirty-five years old who seems like a good person. They're reading a good book. They carry a solid conversation. Give them a recruitment pitch. A couple hundred nights a year, I check into a hotel and interact with the registration desk. Let me share a story. I arrived at a Holiday Inn near Dixon, Illinois (yes, the same city referenced in family nine) at midnight, and it's snowing outside. I walked in, and standing behind the check-in counter was a poised employee with a big smile at midnight.

STRATEGIC HINT FOR YOUR CONSIDERATION

Recruitment is a concerted commitment from everyone in your organization to seek out and engage top-notch people as well as top-notice people who reflect your community.

And you know what she said to me? "You must be Gordon Graham." How'd you know that? "Well, I have one more guest to check-in so I'm assuming you're Gordon Graham." I am. In that one interchange, what does that tell you? She's smart, and she's thinking in advance. "So Mr. Graham, your assigned room has just been painted. I was over there earlier tonight, and it doesn't smell at all of paint. However, to be safe, I left the air conditioner on to remove any possible smell of paint. Please turn the air conditioner off and the heater on when you enter your room." She's a thinker. I'm looking at her desk, and she has books back there and marking pens. She's going to school during the day, and she's working at night. That's the future of my organization. I gave her a recruitment pitch, and she said: "You know, I've never really thought about law enforcement." I said here's my card. CHP's always hiring. We're looking for good people. I think you'd be good at what we do.

She called me a year later. "Mr. Graham, I took your advice. I graduate from the Academy two months from now." Congratulations. Please give me the date, and I'll make sure I'm at your graduation. And she said: "Sir, I didn't attend the CHP Academy. I attended the Chicago Police Department's (CPD) Academy. I joined CPD." Is that good for me? It is because she improves the quality of my profession. Recruit not just for your organization, but recruit for your profession. Everybody needs to be a recruiter. After you recruit people, you have the necessary applicant pool. The second phase is the background investigation process. I'll talk about that process extensively in the next chapter. Here's a hint: Spend your money on background investigations. "Gordon, they're very expensive." I'm a lawyer, and here's my throw down statement: You can pay me now, or you can pay me later. And later is always a lot more.

I'm going to cover background investigations in the next chapter, but I'd like to provide some advanced thoughts. I teach a class on backgrounds and was invited several years ago to talk to the California Background Investigators Association. I started off my class with the following statement: One dollar properly spent on a background can save you one-million dollars in the future. Half the people in the audience have been to my presentations. They've heard me and know me. "There goes Graham. We know what you're trying to say. Backgrounds are important, but there's no way one dollar could save one-million dollars." You don't think one dollar can save one-million dollars? And I would reach into my briefcase and pull out half a dozen settlements and verdicts in excess of one-million dollars that could've been prevented if one dollar was spent on a background check.

STRATEGIC HINT FOR YOUR CONSIDERATION

Had the Massachusetts Department of Public Health Drug Abuse Lab spent one dollar on a background check, they could've prevented one-billion dollars' worth of problems downstream for the State of Massachusetts.

Recently, I was invited back to address this same group. I started off my class with the following statement: Good morning, ladies and gentlemen. Gordon Graham, here. I have you for four hours on the importance of background checks. Let's get started with this assertion: One dollar properly spent on a background check today can save you one-billion dollars in the future. "A billion? No way." You don't believe me? Research Annie Dookhan. It'll prove my point, and I'll provide details on Ms. Dookhan when we arrive at the next chapter. Had the Massachusetts Department of Public Health Drug Abuse Lab spent one dollar on a background check, they could've prevented one-billion dollars' worth of problems downstream for the State of Massachusetts. Backgrounds are an integral tool in getting and keeping good people. They should also be ongoing. More on backgrounds in Chapter 5.

The third phase is the probationary process. I know the following comment varies from state to state and organization to organization, but it's a reliable rule: If you have a probationary process for new employees, it's almost always part of the hiring process. You must terminate women and men on probation who can't or won't do the job. They won't get better over time. "Oh they will." No they won't. They aren't fine wine from Sonoma. They're wine in a box from someplace else. Spend the time to take probations seriously. How are people released from probation? We release them, and we know those employees who are problems lying in wait. Have you ever heard these words in your organization? "How'd he ever get off probation?" We know the employees on probation who are ill-equipped and problems lying in wait. Why do we ignore this reality and release them from probation? Because if you take them on, you get paid X. If you ignore them, you get paid X. So why take them on? Because that's your job. As a supervisor or manager, you must address probation. Don't let people fall through the cracks. It's part of the hiring process. If they can't or won't do the job, then promptly terminate them. Let's examine the probation function a bit further.

STRATEGIC HINT FOR YOUR CONSIDERATION

A probationary period is discretionary time for evaluating the fitness of those you've recruited and offers a forum to compare actual performance against the qualifications of the applicant's résumé.

STRATEGIC HINT FOR YOUR CONSIDERATION

It's prudent to assign a senior leader to be personally responsible for each probationary employee because you'll know where to take appropriate action if the probationary employee ends up causing problems.

Consider a probationary period as discretionary time for evaluating the fitness and suitability of those you've recently recruited. Probationary periods afford you time to think and a forum to compare actual performance, skills, attendance, and personality against the qualifications of the résumé. It also offers an important evaluation period to confirm suitability before matriculating an employee to permanent contract. Employees who are within their probationary period don't qualify for unfair dismissal protection. They are, however, protected against dismissal for protected reasons and wrongful dismissal if you failed to follow any contractual dismissal process. An employer can similarly delay employee benefits until satisfactory completion of the probationary period. The key to a successful probationary period is good performance management and communication of expectations. It'll take time for new recruits to fully understand what's expected of them. As such, you must provide sufficient development time as well as a fair process to address performance and conduct issues. Examples of fair process include scheduling regular meetings to articulate progression and to offer improvement opportunities. As with any other function, a successful probation period requires properly designed, up-to-date, and fully implemented systems. If a rejected probationary employee chooses to challenge their separation formally, the burden of proof falls on them to demonstrate they could perform the responsibilities of the job. The inverse is also true. When you release an employee from probation, the burden of proof shifts to the employer to demonstrate the employee couldn't properly perform the job duties. It's more difficult to terminate employees than it is to reject them while on probation.

Unfortunately some people slip through the cracks and cause problems post probation. What can you do to stop this scenario? First, make sure managers and supervisors understand that probation is part of the hiring process, and if they have probationary employees who can't or won't do the job, they must be rejected. Have you heard these words in your organization: "We'll be shorthanded if we get rid of him." My response: I'd rather be shorthanded forever than have below-grade employees representing our organization. Here's another common quote: "But, we've invested a lot of money in her." Cut your losses early. It'll only become more expensive over time. And another common quote these days: "But, we can't get rid of him because he's a member of a protected class, and we need to increase the diversity of our workforce." Please never think you have to lower standards to increase diversity. That's misguided thinking. There's a multitude of talent in every race, sex, and sexual orientation who'd make excellent water/wastewater professionals. To lower standards to increase diversity is an insult to every other member of that protected class.

STRATEGIC HINT FOR YOUR CONSIDERATION

Too many organizations overrate their people and not because of bias. The real reason is something else: It's easy, and that's a problem lying in wait because the performance evaluation process isn't taken seriously.

So back to the problem: Right now, some probationers are slipping through the cracks because no one in the organization is personally responsible for a given probationary employee. Here's a control measure (system) for you to consider: Assign a senior leader to be personally responsible for each probationary employee. "Mary, you're one of the most respected leaders in our organization. I'm asking you to verify that Peter has the necessary knowledge, skills, and abilities to perform this complex job prior to him being released from probation." This direct assignment of responsibility is effective because when the probationary employee ends up causing problems, we know who to speak to and where to take appropriate action. Keep these processes in mind as it relates to probation.

And finally, the fourth phase in getting and keeping good people is performance evaluations. Chapter 6 will be devoted to this subject, but first a few comments. Take performance evaluations seriously, or do away with them. I'd rather you have no process than a process that's misused and abused. Too many organizations overrate their people. "Why?" Bias. We like our people. How do you honestly evaluate somebody you like? But that's not the real reason. The real reason we overrate people is it's easy. For the executives reading this chapter, how many times has the following scenario happened in your career? Can I help you? "Yes, I just received my performance evaluation, and you have to do something. I've been overrated." How many times has that happened? Never. Nobody complains when they're overrated. If somebody thinks they were underrated, they're pounding on your door. Then you call their supervisor. "I'm hoping you can justify these ratings." Supervisors aren't unwise. They'll frequently rate an insufficient employee as satisfactory because nobody else had ever taken on that employee. And, they may think management won't support them if they evaluate an insufficient employee correctly. Just keep on overrating people. If you take people on, you get paid X. If you overrate people, you get paid X. Why should you take people on and honestly evaluate people? Because that's your job. More on that later in Chapter 6.

Success with getting and keeping good people, pillar one, starts with an unwavering commitment to recruitment, background investigations (initial and ongoing), probation, and performance evaluations. Take these steps seriously. Organizational success, however, requires four other interrelated pillars. We'll now examine getting and keeping good policies (systems), which is pillar two in our five pillars of success.

PILLAR TWO

Policy

STRATEGIC HINT FOR YOUR CONSIDERATION

Well-written policies that are properly derived and annually reviewed by competent counsel impart guidance to employees on how to achieve organizational expectations by doing the right thing, the right way every time.

Good people is pillar one, and the second of the five pillars of success is getting and keeping good policy. Policies (systems) are the mobilization or action component of RPM and the kinetic energy of *real* risk management. Well-designed policies that are kept up to date require identifying your risks and ranking them according to high-risk, low-frequency activities where there's no time to think (core critical tasks) as well as high-risk, low-frequency activities where there's time to think (critical tasks). Risk assessments and risk/frequency matrixes are the investigative tools for the first two components of RPM and must be completed for every job description. You can't avoid tragedies without first recognizing and prioritizing risks. After hiring good people, you must provide up-to-date policies (systems) specific to their job descriptions.

As I travel around the United States, I'm troubled by the lack of good policies. Too often, I consult with organizations that have no policies, out-of-date policies, illegal policies, unconstitutional policies, inconsistent policies, and bad policies. Each of these is a problem lying in wait. Well-written policies that are properly derived and

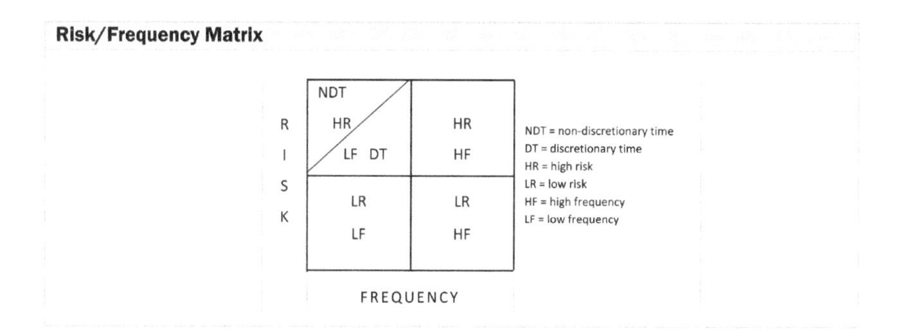

FIGURE 4.2 Risk/frequency matrix. Source: Institute of Safety & Systems Management.

annually reviewed by competent counsel and subject matter experts impart guidance to employees on how to achieve organizational expectations by doing the right thing, the right way every time. That's the purpose of risk assessments and risk/frequency matrixes. These investigative tools provide the base for the pillar of training. The risk/frequency matrix, which is described later in Figure 4.2, is particularly important in prioritizing your training on core critical tasks and critical tasks.

Your people must be trained and fluent on organizational and operational policies. Separate the chafe from the grain by color coding your policy manual. Print your low risk tasks (bottom two boxes) on white paper, print your high risk tasks on yellow paper. Print the core critical tasks (high risk, low frequency, non-discretionary time) on red paper. That way every employee will know the tasks that are overrepresented in tragedy. Attention must be spent on any policy that's high risk in nature. Activities that have the highest probability of causing problems (tragedies) if done incorrectly must be separated from those activities with low consequences if done incorrectly. And that correlates to the importance of risk assessments and risk/frequency matrixes to properly recognize and prioritize activities in every job description that's overrepresented in tragedy. Serious attention must be focused on core critical tasks and, secondarily, on critical tasks. Less focus is necessary on low-risk tasks. Policy manuals, incidentally, are much like the Golden Gate Bridge. They're a work in progress. Whenever someone tells me they finished their policy manual, that's troubling. Consider yourself a temporary custodian who's responsible for maintaining the policy manual during your organizational tenure. Here's a reminder: If I were the lawyer representing an employee who's the focus of your discipline hearing, my first attack would be proof of a written policy that my client allegedly violated. Without a written policy, you'll have great difficulty justifying any discipline.

STRATEGIC HINT FOR YOUR CONSIDERATION

The problem with "shall" in your policy manual is if anything ends up downstream, you've taken all discretion away from the involved employee. Using "should," "can," or "may" offers defensible discretion for your employees.

Let's conclude pillar two with a thought on the unintended consequences of policies where employee discretion is removed. Perform a keyword search on your policy manual for *shall*. When I audit organizations, I always look for *shall* in their electronic policy manual. I have no problem with using *shall* in a policy so long as that's the word you want to use. Perhaps you would be better off using should, can, or may to provide your employees some discretion. If you want to use *shall*, then use *shall*. Here's the problem: If anything ends up downstream, you've taken all discretion away from the involved employee. Again, I don't have a problem with *shall* if that's what you want to use. My law office handled a case for a substantial fire department where they had a policy that said under no circumstances *shall* any fire department vehicle be driven in excess of fifty-five miles per hour (mph). This policy was in place for twenty years without issue. Recently, however, one of their battalion chiefs was driving a Chevy Suburban en route to a training day. It was a seventy-five speed zone. He's driving seventy mph and gets into an accident. Is he in violation of policy? Yes. He's in violation. Don't overuse the word *shall*.

The policy (system) pillar buttresses the people pillar. We'll now discuss training, which propels the five pillars of success. Pillar three ties everything together as long as the training is solid, realistic, ongoing, and verifiable (SROVT).

PILLAR THREE

TRAINING

The most important and underrated pillar within the five pillars of success is pillar three: training. I've already asked you this question: After you were hired and released from probation, when was the next time you had to take a serious test involving preparation? I suspect the answer, besides promotional tests, was never. Let me ask a second question. Have you ever attended a training day where you didn't learn anything? Or worse, where you actually forgot things while you were there? Let me ask a third question. Is your organization more concerned with people knowing things or having a document certifying everyone's completed a training program? And let me ask a fourth question. How do you know what people really know about their core critical tasks or critical tasks prior to their involvement in the activity? The truth is you don't know. The first time you find out if our people don't know their core critical tasks or critical tasks is after the tragedy. By then, it's too late. There has to be a better way. That better way is pillar three (training), which is the foundation for organizational risk management. We have to revisit training and focus it on core critical tasks and critical tasks. "Gordon, you keep referring to these types of activities. What's a core critical task and a critical task?" The former comprise activities that are high risk, low frequency, with no time to think (non-discretionary), whereas the latter involve activities that are high risk, low frequency with time to think (discretionary). Figure 4.2 delves into the details of the risk/frequency matrix.

When I first saw this matrix, I was in graduate school at the Institute of Safety & Systems Management (ISSM). I made a minor change to the top-left box in the 1980s when I was formulating my philosophy on *real* risk management. This change subdivided the top-left box into two areas. Do you see how it's now split between discretionary time and non-discretionary time? The former means you have time to think before you act, whereas the latter means there's no time to think and action is immediate. The upper left corner of the top-left box is the core critical tasks, and they concern me the most. They're the ones that require constant and ongoing training. Training correlates back to each of the other pillars and is necessary to achieve your goal of doing the right thing, the right way every time.

STRATEGIC HINT FOR YOUR CONSIDERATION

The first time you find out if your people don't know their core critical tasks or critical tasks is after a tragedy. By then, it's too late. There has to be a better way, and that better way is daily training.

As a graduate student learning this matrix at ISSM, I had a question: Why didn't the California Highway Patrol explain this chart to me when I was first hired? Why did I have to wait for a specialized course in graduate school to learn something that everybody should know by the age of eighteen? Everything you do in water/wastewater operations. Scratch that. Everything you do in life goes into one of the four boxes in the risk/frequency matrix. Most of what you do involves the bottom-right box: low-risk, high-frequency adventures. Definition: If it goes bad, the consequences are small. The same applies to what you do in the top-right box: high-frequency, high-risk adventures. You do them all the time, but consequences are big if it goes bad. Every now and then, you're involved in the bottom-left box: low-risk, low-frequency adventures. You do them rarely. And occasionally, you're involved in the top-left box that's subdivided: high-risk, low-frequency adventures. Definition: if it goes bad, the consequences are big. The distinction between high-risk, high-frequency activities and high-risk, low-frequency activities is experience. You have memory markers for the former and only training for the latter. The subdivided, upper left box is a problem lying in wait.

What's the real value of a risk/frequency matrix? If you understand this analysis, you can predict with accuracy where mistakes will occur in any job description. That brings us to the value of risk assessments, which are derived from precept one: Your past errors are predictive of future errors. Risk assessments and risk/frequency matrixes are the investigative tools to recognize and prioritize risks. And that correlates to RPM (recognition, prioritization, and mobilization). First, you must recognize the activities for every job description that fall into the top-left box. This step requires risk assessments. Then you must prioritize these activities via risk/frequency matrixes according to their placement in the subdivided, top-left box. Some activities must be performed immediately (non-discretionary time), and some allow you time to think (discretionary time). The upper left portion of the top-left box is the most concerning, as these activities truly afford no time to think. Core critical tasks require regular and ongoing training, as they have a higher probability of getting you in trouble. Training is the mobilization or system component as it relates to the risk/frequency matrix.

STRATEGIC HINT FOR YOUR CONSIDERATION

Core critical tasks comprise activities that are high risk, low frequency, with no time to think (non-discretionary), whereas critical tasks involve activities that are high risk, low frequency, with time to think (discretionary).

Every day must be a training day, and your training should focus on activities that are high risk, low frequency, with no time to think (core critical tasks). Because of their high level of risk, core critical tasks must receive regular training so your people know what to do if they're involved in such an activity. Most activities in the subdivided, top-left box are critical tasks, meaning discretionary time where you can think things through before you act. That may include asking someone who does the activity at a higher frequency how to do it so it gets done right. Most activities are high frequency or discretionary time. Your role is making sure your people in every job description are adequately trained for the non-discretionary time activities (the core critical tasks), and they understand the value of thinking things through when they're involved in a discretionary time activity (critical task).

The risk assessment and risk/frequency matrix processes will confirm that most of your organizational tragedies are caused by employee errors. James Reason (Swiss cheese model) describes these types of errors as being knowledge-based, rule-based, or skill-based. He goes on to examine how errors, lapses, omissions, and mistakes occur in any given workplace. Most of the activities your people perform are high frequency, and their past experience will guide them to do it the right way, the first time. This actuality is derived from recognition-primed decision making (RPD). Gary Klein details the notion in his exceptional book *Sources of Power.* The brain is an accumulation of past experiences or memory markers. Think of it as a hard drive. When the brain gets involved in something it's done many times before, it doesn't have to think it through. It directs current behavior based on past successful behavior. And we stay out of trouble. Give me a good woman and put her in a high-frequency activity, and there's a high probability she'll do it the right way every time. This point was accentuated in a 2001 Scientific American article (paraphrased below) where brain activity of chess players of varying abilities was analyzed:

While test subjects played against computers, researchers studied which parts of their brains experienced gamma-bursts following the computer's move. They found the amateurs' brains exhibited more gamma-bursts in the medial temporal lobe, whereas grandmasters had more gamma-bursts in their frontal and parietal cortices. The researchers concluded the use of frontal cortex by grandmasters, who have memorized thousands of moves, indicates they recognize known problems and retrieve

High Frequency Errors
➲ Complacency
➲ Fatigue
➲ Distraction
➲ Hubris
➲ Risk Homeostatis

FIGURE 4.3 High-frequency errors. Source: Graham Research Consultants, LLC.

solutions from their memories. The use of medial temporal lobe by amateurs suggests these players are analyzing unknown moves and forming new long-term memories.

"Gordon, where do mistakes occur?" Mistakes can and have occurred in each of the four boxes. But the proper question you should ask is slightly modified. In any job description, where are mistakes most likely to occur? And here's the answer: Not in the bottom- or top-right boxes. Rarely do good people, and I assume you have good people, make mistakes on high-frequency events due to RPD. "Gordon, you keep on saying rarely." There are five exceptions, which are illustrated in Figure 4.3, for high-risk, high-frequency activities. Let's review each exception.

STRATEGIC HINT FOR YOUR CONSIDERATION

Most of the activities your people perform are high frequency, and their past experience will guide them to do it the right way, the first time. This actuality is derived from recognition-primed decision making (RPD).

Complacency

I don't care how many times you've dealt with live power, chemicals, or trenching, it's as risky as the first time. The level of risk never changes, but acclamation does change. When high-risk tasks become routine, that's a problem lying in wait. You must address complacency.

Fatigue

I wonder how many tragedies in water/wastewater operations were assigned a proximate cause of X, but the real problem lying in wait was a grossly fatigued employee. If you're not getting seven hours of uninterrupted sleep every night, then you're suffering from fatigue. There's all sorts of fatigue from gross fatigue and circadian fatigue to cumulative fatigue as well as long-term and short-term fatigue. I don't care what you call it. Does fatigue impact decision making, judgment, coordination and balance, disposition, and lifespan? Yes. You must address fatigue in the workplace. It's a big deal. When I study tragedies and read the final reports, rarely do I see any mention of fatigue. How many hours of sleep did the involved employee have in the prior four days? People must know the risk of fatigue. On my recommended reading list is longtime friend, Bryan Vila, who's a national expert on sleep issues in public safety. His technical papers and scholarly articles focus on police officers, but its applicability spans all high-risk professions. Please read his writings, and take fatigue seriously.

STRATEGIC HINT FOR YOUR CONSIDERATION

When I study tragedies and read the final reports, rarely do I see any mention of fatigue. How many hours of sleep did the involved employee have in the prior four days? People must know the risk of fatigue.

Distractions

Don't become distracted while performing high-risk activities. Cellphones are a distraction, as are in-car computers. I write a monthly piece for the New York State Association of Chiefs of Police Magazine, and one of my pieces was on workplace distractions. The California Highway Patrol used to have motorcycle cops on Sunset Boulevard, and these cops were grossly overrepresented in rear-ending stopped traffic. They were distracted. "Well, Gordon, it had to be the beautiful people walking up and down the street." No, it wasn't. It was the plate-glass windows on the storefronts where the motorcycle cops were able to look at themselves while they rode down the straight. "Wow, I look good," and they were rear-ending people. I'm not making that up. Don't allow people to get distracted while they're doing high-risk activities. Distractions are preventable, but it requires communication and training to prevent such behavior.

Hubris

I love confident personnel, but cocky personnel are problems lying in wait. When people start thinking they've been doing their job so long that nothing bad is going to happen, something bad is going to happen. Don't get cocky.

Risk Homeostasis

And finally item five is a fascinating phenomenon known as risk homeostasis. "Gordon what's that all about?" That's where we do things with the goal of making people more safe, but in reality we make them less safe. "Gordon that doesn't make sense." That's why it's a phenomenon. Talk to old firefighters. They'll tell you the following about intuitive exits: "Gordon, your ears will tell you when to get out of a burning building." You could hear something, chief? "No, they get so hot that you had to leave." Firefighters' ears used to get so hot they had to exit buildings. Well, we don't want firefighters getting their ears burned so we gave them what? Hoods. So what do they do now? They go in deeper, stay in longer, and buildings collapse on firefighters. Have hoods made firefighters more safe or less safe?

STRATEGIC HINT FOR YOUR CONSIDERATION

Cocky personnel are problems lying in wait. When people start thinking they've been doing their job so long that nothing bad is going to happen, something bad is going to happen.

Nineteen firefighters were killed in Arizona. Remember that terrible tragedy. What was the big conversation after that tragedy? How could we improve the quality of fire shelters? If you've been reading any of my musings over the years, I've been railing against fire shelters. Canadian wildland firefighters don't carry fire shelters, and neither do Russian, Greek, Australian, or New Zealand wildland firefighters. All those countries teach their wildland firefighters to exit if they encounter an uncontrolled wildfire. Have fire shelters made people more safe or less safe? The California Highway Patrol bought BMW bikes in the 1990s. It was the only bike you could buy with anti-lock brakes. You know what? Better brakes will avoid some crashes. Collisions didn't go down. They went up. Once cops figured out they had better brakes, they started driving faster. That's the phenomena called risk homeostasis.

Let's get back to our discussion on risk/frequency matrixes. Most of what you do, you're doing right because most of what you do, you've done before. Address complacency, fatigue, hubris, distraction, and risk homeostasis, and you can stay out of trouble. Rarely do we make mistakes on high-frequency activities. Mistakes are more likely to occur on low-frequency activities. And I'm not concerned about low-risk, low-frequency activities on the bottom-left box because those activities are low risk. Even if it goes bad, the consequences are small. I'm concerned for water/wastewater personnel in the top-left box. When good people, and again I believe you have them, perform high-risk, low-frequency activities, they're more likely to make an error. That bears repeating: In any profession, people are more likely to make mistakes in the top-left box than anyplace else. Now to be fair, not every mistake is going to end up in a tragedy. But if you make enough mistakes, sooner or later, all the holes in the Swiss cheese will line up, and tragedy occurs. Let's review some examples.

Do you remember Asiana Flight 214 landing at San Francisco International Airport? The aircraft was a Boeing 777 arriving from China and missed the runway by about fifty feet. It broke in half, and people ended up dead. I read the final report. How many hours of flight time did the pilot have on the Boeing 777? I think the number was forty-three, of which thirty were in a simulator. That's only thirteen hours of actual flight time. How long was the flight from China? Thirteen hours. That activity was high risk, no frequency. One passenger was also killed when firefighters ran over her. Like the pilot, those firefighters were playing in the top-left box: high risk, low frequency. Not one of them had been involved in a commercial aviation crash in their career: high risk, low frequency.

STRATEGIC HINT FOR YOUR CONSIDERATION

Risk homeostasis is a phenomenon where we do things with the goal of making people more safe, but in reality we make them less safe. What are some examples of risk homeostasis in your profession?

Speaking of firefighters, who dies fighting wildland fires? Overrepresented are municipals, seasonals, and volunteers? Why? For a career wildland firefighter, fighting a wildfire is high risk, high frequency: top-right box. They do it all the time. They've memory markers, behavioral scripts, and experience. Conversely, place a smoke jumper from the Forest Service in a high-rise fire, and she's going to get in trouble: high risk, low frequency. When I was a new cop, they gave me a revolver. Fifteen years into my career they took the revolver away and gave me a semi-automatic pistol. Every police department that transitions from wheel guns to semi-automatic pistols in the first year has a higher number of unintentional discharges. Why? Because for a while, their people are playing in the top-left box: high risk, low frequency. You give your people a new truck with a higher center of gravity, and for a while, they're playing in that top-left box: high risk, low frequency. You give them a new chainsaw: high-risk, low-frequency box. Whenever possible, stay out of that top-left box.

Mrs. Graham loves going to Scotland. Our daughter did her graduate work there, and we used to visit regularly. So one night, Mrs. Graham and I were watching Wheel of Fortune, and the prize puzzle was a trip to Scotland. And here comes the tears. "Poor Sarah. She's over there all by herself, and we haven't seen her in almost two weeks. I think I'll fly over tomorrow and visit her." Sweetheart, I won't be able to go. I'm working tomorrow. "That's okay. I know how to get there." How are you going to get there? "Same way we always do. I'll fly United into Edinburgh, rent a Hertz car, and drive up to Saint Andrews." Sweetheart, how many times have we been over there? "I don't know? Maybe twenty." Have you ever seen me rent a Hertz car? "No, you always hire off-duty cops to do all the driving while we're in Europe." Yes, *real* risk management isn't just a vocation. It's a way of life. The moment you leave the Hertz lot, you're playing in the top-left box. You're driving on the wrong side of the road, and for those who think that's not a big deal, wait until you come to a roundabout. Every day, a US citizen is getting into trouble on a roundabout. Why? It's high risk, low frequency. When the Brits come over here, they get in trouble. They step off the curb and look which way? To the right, and our cars come to the left. It's all about the risk/frequency matrix. High-risk, low-frequency activities (core critical tasks and critical tasks) concern me.

Let's discuss solid, realistic, ongoing, and verifiable training, otherwise known as SROVT. This program encapsulates my life's work and the reason for the award

I'm most proud of receiving: the excellence in law enforcement by Governor Pete Wilson. This award was based on the risk/frequency matrix from ISSM. As students, we had guest lecturers coming in talking about military safety, and one lecturer talked about submarine safety and submarine fires. It's a big deal to have fire on a submarine, and it must be extinguished immediately to avoid crew death. The Navy identified and evaluated the risk and put together an effective, yet simple, control measure (system): Every day, every submariner is trained on firefighting. Do you know that most submariners will go through their entire career without experiencing an onboard boat fire? They still train every day.

STRATEGIC HINT FOR YOUR CONSIDERATION

Amanda Ripley aptly sums up the need for daily training in her book, The Unthinkable: "These unthinkable events, these things that aren't supposed to happen have a nasty habit of popping up every now and then."

This lecturer then started talking about undersea collisions, which are an equally big deal as fires. Undersea collisions can rip the front end of a submarine. And if you don't secure the hatch quickly and properly adjust the buoyancy throughout the entire boat, the boat will creep and enter crush depth. So what did the Navy do? Every day, every submariner is trained on damage control and response to undersea collisions. Why? Most submariners go through their entire career without experiencing an undersea collision. Why do they keep on training? On my recommended reading list is a great book by Amanda Ripley titled *The Unthinkable*. She sums it up very nicely. "These unthinkable events, these things that aren't supposed to happen have a nasty habit of popping up every now and then." Please research the USS *San Francisco* January 8, 2005. The USS *San Francisco*, a Los Angeles-class nuclear submarine was making a speed run from Guam to Brisbane. It went off course and collided into a seamount. The impact was violent: flank speed into a seamount, five hundred feet below sea level. The collision compromised the front end of the San Francisco. There were one-hundred-thirty-seven sailors on the boat that night. How many should be dead? All of them. How many died? One machinist mate was crushed on the initial collision. He was working in the bow. Everyone else survived. They had zero memory markers created by frequency, but they had memory markers from constant and ongoing training.

STRATEGIC HINT FOR YOUR CONSIDERATION

To avoid errors in core critical tasks, your organization must implement a solid, realistic, ongoing, and verifiable training (SROVT) program that's repetitive and deployed pre-incident and not post-accident.

Every day is a training day, and every day we must train. I built that program into the California Highway Patrol back in the 1980s, and it was the basis for the award bestowed by Governor Pete Wilson. Every day is a training day. I've also built a company for public safety organizations where every day is a training day. I'm committed to doing the same for water-related entities, which is standardizing best practices and buttressing these policies with constant and ongoing training. Every day must be a training day for your organization. "Gordon, I used to watch CHIPs, and you guys sat around talking for hours. You don't know the way water-related entities work. We don't have a daily briefing. We get in our truck or work area and start our job. We don't have time for daily training." Really? At some time during the day, do your people log on to their computers? I bet most of your people do that. When people log on at the start of the day, why isn't the first thing on their screen a training bulletin specific to a core critical task and followed up with a test question? There must be training and testing every day on core critical tasks.

Let's delve deeper into the core characteristics of initial and ongoing training. To avoid errors in activities that are high risk, low frequency, with no time to think (non-discretionary), you must have solid, realistic, ongoing, and verifiable training (SROVT). That equates to repetitive training up front before the tragedy. Repetitive training after the fact doesn't work. That's classic lawyer-think. You're responsible for training. Every day must be a training day, and the training should focus on activities that are high risk, low frequency, with no time to think, as well as those activities that are high risk, low frequency, with time to think. And what do we call such activities? Core critical tasks (non-discretionary time) and critical tasks (discretionary time). Playing the what-if game in training scenarios works.

Consistent with your good organizational policy, there should be pre-incident knowledge level verification. Remember precept two: There's always a better way to prevent tragedies through improved systems. How close can we bring pre-incident training to the actual activity? It should be as close to the actual life test as possible. Your training on core critical tasks must be properly documented. Training records are essential for core critical tasks because without documentation, many will assert the training didn't exist. That's particularly important when you're downstream in litigation.

STRATEGIC HINT FOR YOUR CONSIDERATION

Your people only have training and experience from which to rely. If you don't have experience, then you must rely on training. If your training was some time ago, that's a problem lying in wait.

STRATEGIC HINT FOR YOUR CONSIDERATION

When people log on at the start of the day, the first thing on their screen should be a training bulletin specific to a core critical task and followed up with a test question. There must be training and testing every day on core critical tasks.

Virtually all water-related personnel start their career with training. Your hard drive has been partially loaded through training. After you complete your initial training during the probation phase, when's the next time you had to take a serious test that required advanced preparation? For many of you, the only test you ever had to take was the activity itself. That's the first time you're presented with a situation posing a very specific question requiring a very specific answer. Where's the pre-incident knowledge level verification for core critical tasks? How do you know what your people know about these high-risk, low-frequency, with no time to think (non-discretionary) activities prior to their involvement in the activity? The answer is simple: You don't know. Your people only have training and experience from which to rely. If you don't have experience, then you must rely on your training. If your training was some time ago, that's a problem lying in wait, and your goal of proper and safe conduct is substantially diminished. Training isn't the sole responsibility of supervisors. It's the responsibility of everyone. No one loses with highly trained, highly qualified professionals responding to and handling assigned tasks, incidents, and events (activities). Instilling that type of organizational culture starts with your executives, but it must be reinforced by your managers and supervisors. It requires buy-in from everyone.

Water/wastewater personnel start off their career with solid and realistic training (S&R). I've no objection with initial training in most professions throughout our great country. I've seen your initial training and am impressed with it. Where many organizations fail is in providing ongoing and verifiable training (O&V). Ongoing means regularly and in consumable portions. Verifiable means making sure you know the law (if any), the organizational policy, and the appropriate technique prior to your involvement in the activity. How is a SROVT program implemented? It involves a four-step approach to success. First, you must identify the core critical tasks in each job description within your organization. That's done by codifying the total activities and removing activities where there's low risk, high frequency, substantial experience, and discretionary time. That leaves you with those activities that are high risk, low frequency, with no time to think (non-discretionary). That was exactly what Richard Rescorla was doing for the employees of Morgan Stanley. He was preparing them for a core critical task. It's what flight attendants do before takeoff. They're preparing passengers for a core critical task. I do the same thing at hotels when I count my steps to the emergency exit and prepare my shower for the next morning. I'm priming myself for a core critical task.

It's now time to review your up-to-date job descriptions with a marking pen. Use the above formula (i.e., total activities from a job description minus activities that are low risk, high frequency, substantial experience, and discretionary time) to identify the activities that fall in the subdivided upper portion, top-left box of the risk/frequency matrix. Every water/wastewater professional has different core critical tasks and critical tasks. The job description of a treatment operator is different than a distribution operator or an administration employee. Supervisors are responsible to know their own core critical tasks and critical tasks as well as those of their subordinates. I recommend a regular assessment tool to identify areas of focus so you can ascertain what your employees really know and don't know. The performance

Initial Training Guidelines

- Identify the hazards and assess the risks involved in this training activity.
- Analyze available control measures (systems) applicable to this training activity.
- Inform trainees of the involved control measures (systems), and why they're important.
- Implement the control measures (systems) throughout the involved training activity.
- If rules aren't followed, address the deviations appropriately.

FIGURE 4.4 Initial training guidelines. Source: Graham Research Consultants, LLC.

evaluation process could be used more effectively in verifying knowledge, skills, and abilities, particularly in the area of core critical tasks and critical tasks.

Second, after identifying the core critical tasks and critical tasks, make sure you have a policy that addresses the identified activity and fully incorporates applicable law and required technique. Sometimes, the development of an SROVT program will find core critical tasks and critical tasks that have no policy. Third, after identifying these activities and finding the right way to perform the activity (following law, policy, and technique), make sure your people have solid and realistic training (S&R). Do they really know how to perform these activities? In most states, that's usually achieved during initial training and probation phase. It's called the immersion process, in which we spend a lot of time covering the identified activity and making sure our personnel know how to do the activity. Please make sure that's happening at the start of one's career. With respect to initial training involving live practice, Figure 4.4 depicts a five-step risk management approach for the S&R component of this process.

Fourth, after they really know it, follow up with ongoing and verifiable training (O&V). The high level of knowledge obtained during the immersion process will deteriorate over time unless spiked by frequency or, in the absence of frequency, by additional O&V training. An analogy would be booster shots. We vaccinate our children, but that's not good enough. We have to occasionally give them a booster to maximize the protection. With respect to ongoing training, once or twice a month per activity is adequate. Ideally, each water/wastewater professional should receive a six-minute training session per day focusing on these high-risk, low-frequency activities in general and the non-discretionary time activities in particular.

STRATEGIC HINT FOR YOUR CONSIDERATION

Why six minutes of daily training? Six minutes times a five-day workweek is one-half hour per week, two hours per month, and twenty-four hours per year. That's a lot of training hours, and it costs you nothing.

Why six minutes? Do the math. Six minutes times a five-day workweek is one-half hour per week, two hours per month, and twenty-four hours per year. That's a lot of training hours, and how much did it cost you? Nothing. Adults learn better by repetition than by immersion. Have you ever had a training day after working night shift? Did you leave that training day with any real learning? Pre-incident knowledge level verification with a one-hundred percent score is the goal. Random audits and inspections will assist you in this regard. You now have a SROVT program and are en route to your goal of proper conduct for every organizational activity. There are three benefits to having a SROVT program in place. When your employees do the right thing, the right way every time, your personnel and public are protected, downstream exposure is reduced, and loyal customers are formed.

Your take back to work assignment is to identify those high-risk, low-frequency activities in every job description including front desk, administration, clerical, billing, accounting, maintenance, distribution, and treatment. You must identify those high-risk, low-frequency activities. "Well, Gordon, you're not going to tell us?" I don't know. I've never done your job. "How can we find out?" Remember the three precepts of risk management: (1) There's no new ways to get in trouble; (2) there's always new ways to prevent trouble; and (3) predictable is preventable. Let's bring this back to your work assignment. If you want to identify the activities in every job description that end up in that top-left box, you must study your past tragedies and near misses. The errors you're going to make can be predicted from the errors already made. There's no new ways to get in trouble. If you want to find out how front-desk people are going to get in trouble, you need to study how they've already been in trouble. If you want to know how chemists are going to get in trouble, you need to study how they've already been in trouble. Study your past tragedies.

"Gordon, we've never had an employee at the front desk get in trouble." Just because it hasn't happen to you, doesn't mean it hasn't happened. Study similar water-related entities as well as industry trends and ask the same question. That's the benefit of affiliating with state and national water/wastewater associations. They have data from around the country on all types of water-related entities. Ask them: In this job description, how do people end up downstream in that top-left box? "Well, Gordon, you can't tell us what they are. Can you tell us how many there are?" Yes, five percent. If I had to hazard a guess, only five percent of what you and your people do falls in that upper left side of the subdivided top-left box. Of this five percent, approximately two percent falls into the high risk, low frequency, non-discretionary time category. Ninety-five percent of what your people do is either high frequency or low risk or discretionary time. The upper left side of the subdivided top-left box is five percent. "Well, Gordon, assuming we do a thousand activities, that means there's about fifty activities in the upper level portion of the subdivided box?" Yes, about fifty. "Gordon, that's still a pretty big number." That's why you need to prioritize them, which goes back to the concept of RPM (recognition, prioritization, and mobilization).

STRATEGIC HINT FOR YOUR CONSIDERATION

Malcolm Gladwell and his great book, Outliers, confirm it takes ten-thousand hours to master any profession. You must deploy a ten-thousand-hour rule for applicants to promote to supervisor or manager.

PILLAR FOUR

Supervision

Good people I believe you have. Good policy I'm pretty sure you have. Effective training requires a commitment to a SROVT program where every day is a training day with a focus on core critical tasks and critical tasks. The fourth pillar of success is supervision, and we'll devote an entire chapter to this important topic. Let's briefly discuss some key points. I already shared the quote from Commander Ross Swope:

> On any given day, nobody knows where the general manager is, and nobody knows where the department manager is, but everybody knows where their supervisor is. And many employees will modify their behavior based on which supervisor is on duty.

When you study tragedies in the workplace, all too often it comes down to supervisors not behaving like supervisors, or supervisors who tried to behave like supervisors but received no support from management. "Gordon if you were the executive-in-charge, how would you do it?" Thought number one. I like the idea of ten-thousand hours. That's right: ten-thousand hours. Malcolm Gladwell is all over my recommended reading list, and he's got a bunch of great books out there, including one called *Outliers*. I read his ten-thousand-hour rule in *Outliers*. In order to master any profession, it is recommended you do it for at least ten-thousand hours. He then talks about Kobe Bryant (1978–2020), Itzhak Perlman, Michelle Kwan, and Tiger Woods. All these people achieved the pinnacle of their profession through hour-after-hour-after-hour of practice. Ten-thousand hours.

STRATEGIC HINT FOR YOUR CONSIDERATION

For discipline to work, it must be prompt, consistent, fair, and impartial. Discipline must never be a function of consequence but a function of policy. All's well that ends well is a problem lying in wait.

The primary mission of a supervisor is systems enforcement. Someone has to enforce your organizational policies and procedures (systems). Genuine concern for your personnel is paramount. Be out and about and catch your people doing something right and take the time to document it. The number one complaint I receive from line-personnel (around the world and in every profession) is "the only time I hear from my supervisor is when something is wrong." Please think this one through and make it your goal to document observed good behavior regularly.

PILLAR FIVE

Discipline

The last of these five pillars of success is the discipline component. And I've already touched on it. Discipline isn't a function of consequence, or all's well that ends well.

"Did anybody get hurt?" No. "Did any property get destroyed?" No. "All's well that ends well." You can have an activity like trench shoring end up okay in which there are major violations of safety rules. I refer back to trench shoring because many construction workers die in trench operations, and this activity occurs in water/wastewater operations. "We didn't properly have it shored up, but nobody was hurt so all's well that ends well." Just because it ended up okay doesn't mean we can ignore it. You can't rely on luck. You must rely on process (systems).

Discipline must be prompt, consistent, fair, and impartial. It's never a function of consequence, but a function of process. When people don't follow process, it must be addressed, notwithstanding outcome. Take the time to slow down. Take the time to think it through. If possible, contact human resources personnel. Ask them how to handle it. You just can't fire people in most states. There's procedural due process. Discipline is important. People make mistakes, and they need retraining. That's okay. When people cross the bright line of ethics and integrity, however, that's not okay. You must terminate them. So there you have it, folks. The five pillars of success: people, policy, training, supervision, and discipline.

FINAL COMMENTS

STRATEGIC HINT FOR YOUR CONSIDERATION

Everyone benefits from having a sound, professional work environment. Remember, most things that go wrong in life are highly predictable, and if predictable, it's preventable.

If you take the time to study the consequences (tragedies) that occur in water/wastewater operations, almost without exception you can link them to the root causational factors of people, policy, training, supervision, and discipline. Sometimes, it's only one of these, sometimes more than one, and sometimes all five of these root causes are involved in a single activity that ends up in tragedy. While there are thousands of proximate causes (the event that instantly preceded the tragedy), there are fewer root causes. High-reliability organizations are underrepresented in tragedies and overrepresented in employee fitness and customer loyalty because they're focused on getting and keeping good people, building good policy, assuring great initial and ongoing training, providing adequate supervision, and having a viable discipline process in place to address employees who don't follow organizational policy. Take away one of these pillars and bad things will happen. They're inextricably intertwined in supporting your mission and achieving success. Everyone benefits from having a sound, professional work environment. Remember, most things that go wrong in life are highly predictable, and if predictable, it's preventable.

And let me wrap up this chapter with three questions. Question one: Why do most things go right? Because you have good people, and you're putting them in high-frequency tasks. Their recognition-primed decision making (RPD) kicks in, things go right, and they avoid tragedies. Questions two: Why do things go wrong? Two reasons. Somebody did something bad on purpose, or somebody made a mistake. The intentional misconduct from employees can be addressed if we do a better job at recruitment, backgrounds, probation, and performance evaluations. The negligent conduct from employees can be taken care of if we did a better job at training. Which brings us to question three: How do we train? Training must be more than a document certifying somebody went to a class one time, in time, some time ago. For training to be successful, it must be focused on core critical tasks and, secondarily, critical tasks. Every day has to be a training day.

I look forward to reconnecting in Chapter 5 where we're going to discuss the importance of background investigations. Until then work safely.

Chapter Takeaway on Media Relations/Crisis Communication

Please refer to the media relations/crisis communication material in the Addendum section and ask yourself how you would address the following scenario:

Water main breaks are an expected and budgeted exposure. You know that, and you've made sure your employees are properly trained to quickly and safely stop water flow as well as restore service. Your organization's solid, realistic, ongoing, and verifiable training (SROVT) program on main breaks is best-in-class. Each of your line-personnel has a one-hundred percent knowledge level verification for this activity. They know what to do, and they do it well. There's just one problem. You haven't offered ongoing training when breaks occur under a thoroughfare. That's never been a priority because almost all your water mains were installed under low-volume roads. This afternoon's break ended the good luck streak. Your employees quickly arrived on scene to fix the break. Their training naturally kicked in except for traffic control at the worksite.

You regret not including ongoing training for traffic control since it clearly meets the definition of a core critical task. Training for this activity is only covered during the initial probation phase, which means pre-incident knowledge level verification is low. This problem lying in wait has finally surfaced. Your work crew closed off the lane adjacent to the break but didn't create enough notice for drivers to merge. Traffic cones protruded into the other lane and created more confusion. Shortly after implementing these measures, a massive collision involving major injuries and several vehicles occurred. Police and first responders are on scene. So is the local news and print media. It's clear the worksite diversion protocol was insufficient. Phone calls to you for commentary are now coming through the switchboard. Reporters are arriving outside your headquarters to supplement their on-scene video and are requesting an interview for the evening news.

This accident is a tragedy involving public injuries and organizational embarrassment as well as likely civil liability. How do you answer the inevitable questions on training and policy? And how do you balance the need for honesty and compassion with transparency? You know litigation is inevitable, but you must protect your organizational goodwill. You can't hide from the press, but you must be careful because you know the bias of reporters in how they will portray you and your organization. What do you do?

5 Comprehensive Background Investigations

Past Behavior Is Predictive of Future Behavior

2019 Photo Courtesy of San Diego County Water Authority

STRATEGIC HINT FOR YOUR CONSIDERATION

There will never be a complete elimination of intentional misconduct (internal), but comprehensive background investigations, coupled with the three other interrelated control measures, have proven to be highly effective.

DOI: 10.1201/9781003229087-5

SUMMARY

Comprehensive background investigations are a validation tool for your hiring practices and play an integral role in screening bad actors who have an inclination for intentional misconduct. This tool identifies relevant behaviors in an applicant's past that suggest a sufficient or insufficient level of integrity to meet the needs of your organization.

Properly screening recruits from your applicant pool necessitates executive support, resilient systems, and sufficient resources. An integral part of the screening process includes a coherent sequence of rules for improving the process and ensuring its proper application.

Chapter discussion will conclude with recommended control measures for ongoing background investigations. This possibility must be managed, as good employees occasionally turn bad and can cause a variety of organizational tragedy.

Gordon Graham here, and welcome back to our journey toward *real* risk management. The first four chapters laid the foundation for my philosophy and its various components. We know high-reliability organizations, which embody *real* risk management, have an inexorable pursuit to do the right thing, the right way every time through self-reinforcing systems and culture, with discipline serving as its adhesion and continual improvement representing its organizational purpose.

Importance was similarly conveyed on a process known as RPM (recognition, prioritization, and mobilization), which involves recognizing and prioritizing risks through the ten families classification system and its investigative tools of risk assessments and risk/frequency matrixes. These tools ensure the effective mobilization of resilient systems that are properly designed, up to date, and fully implemented.

A checklist for success was similarly presented to maximize accuracy and consistency by applying a methodical process of risk, systems, service, accountability, and integrity. The derivative of these themes include fulfillment of organizational mission as well as underrepresentation in tragedies and overrepresentation in employee fitness and customer loyalty.

The book's apogee is organizational risk management, where theory is applied to practice. Organizational risk management is the process of identifying and addressing your precursors to tragedy, known as root causes or problems lying in wait. These root causational factors comprise people, policy, training, supervision, and discipline. The conversion of your problem factors into solution factors creates pillars of success.

This theory is yielded from the three precepts of *real* risk management: (1) Past tragedies are predictive of future tragedies; (2) there are always ways to reduce tragedies by improving systems; and (3) identifiable risks are manageable risks. The preceding precepts confirm that most tragedies involve negligent conduct from

Getting & Keeping Good People
Four Interconnected Factors
➲ Recruitment
➲ Background Investigations
➲ Probation
➲ Performance Evaluations

FIGURE 5.1 Getting and keeping good people. Source: Graham Research Consultants, LLC.

otherwise good employees who make mistakes while performing their activities. Sometimes, tragedies aren't caused by negligent conduct but rather intentional misconduct from bad actors in your organization. The latter will be the discussion of this chapter, "Comprehensive Background Investigations," and Figure 5.1 illustrates the four interconnected elements of pillar one: getting and keeping good people (recruitment, background investigations, probation periods, and performance evaluations).

The chapters that follow will delve into the five pillars of success (people, policy, training, supervision, and discipline), which comprise organizational risk management. The penultimate chapter will address job-based harassment and its abolition through a policy of zero tolerance that's enveloped in the other four pillars. Our final chapter will encapsulate an ethical and a sound decision making process so your employees are taught the importance of thinking before engaging in any high-risk activity where there's time to think. We'll also be introducing logical and coherent rules for the remaining chapters. These rules comprise simple, but not simplistic, techniques to improve employee performance by providing clarity so activity-specific issues are addressed properly and consistently throughout your organization. Let's begin by examining the rules for and necessity of comprehensive background investigations.

Every profession has bad actors, and sooner or later they're caught, fired, or prosecuted. Such intentional misconduct must be taken seriously and can be managed by developing, selecting, and implementing control measures (systems) up front to reduce the probability of harmful consequences (tragedies). Systems like good recruitment, comprehensive background investigations, understanding the probationary period, good supervision through performance evaluations, and consistent discipline can minimize the exposure. To be sure, there will never be a complete elimination of intentional misconduct (internal), but these four interconnected control measures have proven to be highly effective. We'll delve deeply into the purpose of comprehensive background investigations as well as the risk management techniques to deliver a proper end result.

Comprehensive background investigations are a validation tool for your hiring practices, and this latter process begins with recruitment. To be effective, recruitment requires high standards and active promotion from your employees. Lowering standards or having an applicant puddle are problems lying in wait. Proper recruitment involves patience and detail to properly evaluate candidates and filter out those who are unfit or unable to do the complex jobs in water/wastewater operations. You can test for intelligence, strength, agility, and flexibility. You can also test for vision and body mass index. You can't test for integrity. Anyone who tells you there's a test for integrity is misguided. The only way to test for integrity is to recognize that the best predictor of future behavior is past behavior. If you hire people who have a history of behavior that demonstrates a lack of integrity, you have a problem lying in wait.

STRATEGIC HINT FOR YOUR CONSIDERATION

You can prudently spend a reasonable amount of money up front to filter out bad actors, or you can spend a lot more money downstream to cover the liability costs when a bad actor in your employ does something unprincipled.

The checking of past behavior is the crux of the background investigation process. Its purpose is to find relevant behaviors in the applicant's past that will demonstrate a sufficient level of integrity to meet the needs of the water-related entity. You have to make sure you're hiring good people. Sadly, as I travel around the United States and talk about the importance of comprehensive background investigations, I learn too many organizations stop with a criminal history check. If you don't learn anything else from this chapter, please remember the following: The difference between having a criminal history and not having one comes down to being caught. Nineteen terrorists in 2001 brought down four planes and caused massive carnage to our great country. None had a criminal history in the United States prior to their nefarious actions on September 11.

A comprehensive background investigation for every applicant is essential, and, yes, this activity will cost money. You can prudently spend a reasonable amount of money up front to filter out bad actors, or you can spend a lot more money downstream to cover the liability costs when a bad actor in your employ does something unprincipled. And here's my throw down statement as a lawyer: You can pay me now, or you can pay me a lot more later. There are cases throughout the United States of sexual molestation and lewd behavior by cops, firefighters, coaches, and yes, water/wastewater professionals, who had a history of inappropriate behavior but were hired by another organization. It isn't uncommon for sexual predators to drift from one organization to another without anyone aware of their illegal and immoral behavior. Similarly, I can point to numerous water/wastewater personnel who were fired for bad driving but hired by another organization post termination.

STRATEGIC HINT FOR YOUR CONSIDERATION

Your background investigation process must include validated information sources that can be compared against the applicant's personal history statement (PHS).

In my classes on background investigations, I provide startling examples of lax hiring processes that allowed sexual predators, thieves, murderers, bullies, madcaps, and other bad actors into water-related entities and governmental organizations. Your profession isn't some evil cauldron that hires good people and turns them into bad people. Occasionally, you hire bad people and put them in a position of public trust, and they intentionally do bad things in your employ. Your background investigation process must include the selection of qualified investigators and the development of appropriate and legal application documents to mitigate this reality. It should similarly comprise the identification and utilization of validated information sources and processes to analyze this data against the applicant's personal history statement (PHS). You don't have to reinvent the wheel. I always recommend you review the background investigation process of your state police. You can learn a lot from their process, and, along with your competent legal counsel, you can customize a process that works for your organization.

Quality and compliant information is crucial to enable organizations to hire the right candidates. There's no perfect background check or single database that includes all civil and criminal records. Every organization, however, can control the due diligence and quality assurance standards that comprise their background screening program. This activity is high risk, but it's also discretionary time. You must consult with your competent counsel regarding the selection of a backgrounds vendor and the type of reports you'll be requesting. The summary that follows encompasses material points and criteria to consider when selecting vendors and reports for this important function.

Vendor Selection

NAPBS® accreditation (National Association of Professional Background Screeners and Background Screening Credentialing Council): NAPBS® provides a singular, cohesive industry standard via the Background Screening Agency Accreditation Program (BSAAP). To become accredited, a consumer reporting agency (CRA) must pass an onsite audit that's conducted by an independent auditing firm of its policies and procedures. This accreditation is important because screening reports are categorized as consumer reports and regulated at the federal and state level. *Consumer protection servers housed offsite in secure tier-four facilities*: Criteria to consider include information security policy, data security intrusion, detection and response, stored data security, password protocol, electronic access control, physical security, consumer information privacy policy, unauthorized browsing, record destruction, consumer disputes, sensitive data masking, and database criminal reports. *In-house compliance administrators*: Criteria to consider include designated compliance person(s) as well as operational conformance with state consumer reporting laws, driver privacy protection act (DPPA), state implemented DPPA, prescribed notices, and regulatory conformance with Fair Credit Report Act (FCRA), Equal Employment Opportunity Commission (EEOC), and Federal Trade Commission (FTC). *Quality business practices*: Standards to consider for acceptable organizational character include insurance, client credentialing, vendor credentialing, consumer credentialing, document management, employee certification, worker training, visitor security, employee criminal history, quality assurance, and certification. *Trusted federal, state, and local government partner*: Criteria to consider include long-standing relationships with regulated entities in federal, state, and local government.

STRATEGIC HINT FOR YOUR CONSIDERATION

The background investigation process is high risk but also discretionary time and requires consultation with your competent counsel in selecting a vendor as well as the type of reports you'll be requesting.

Recommended Reports

STRATEGIC HINT FOR YOUR CONSIDERATION

A social security number (SSN) trace includes validation of the SSN from the Social Security Administration (SSA) as well as state of issuance and date range. It also identifies aliases, additional names, and address history associated with the SSN.

Social security number (SSN) trace: An address history search encompassing credit header files, national databases, real property records, voter registration, and other proprietary sources. SSN trace includes validation of the SSN from the Social Security Administration (SSA) as well as state of issuance and date range. It also identifies aliases, additional names, and address history associated with the SSN. *National sex offender and national security report*: A comprehensive criminal report accessing thirteen-hundred different criminal security sources and over five-hundred-fifty-million records nationwide from all available, regularly updated criminal sources. This compilation searches booking records, felony and misdemeanor criminal convictions, prison parole, and probation information obtained from county and municipal courts, state departments of corrections, state and county criminal records repositories, sex offender registries, and other criminal justice agencies, including twenty national and international security sources. *Driving records check*: Encompasses physical description, type of license issued, moving violations, collisions, and driving under the influence convictions. *Employment credit report*: Comprehensive snapshot of a candidate's financial history. It offers details of an individual's debt obligations, payment histories, and public record information including bankruptcies, liens, and judgments. Instant turnaround time. *County criminal records check*: Provides a county court search for felony and misdemeanor criminal conviction information for the previous ten years. Checks are performed in counties where the candidate has lived for the past ten years. *Federal criminal records check*: Court search for federal criminal conviction information for the previous ten years if available. *Education verification*: Confirms highest degree of reported educational background. *Employment verification*: Validates ten years of employment history including dates of employment, position held, and salary (if available).

The above-referenced criteria for vendor selection and recommended reports underscore the importance of detail and thoroughness. You can't expedite or abridge the background investigation process, and the ensuing example accentuates that point. I mentioned some advanced thoughts on Annie Dookhan in Chapter 5. For those of you who didn't research her, Annie Dookhan was a 2002 university graduate who earned a Bachelor of Science in chemistry. She wanted to be a chemist in a crime lab and was hired by the Massachusetts Department of Public Health in their drug abuse lab. The State of Massachusetts completed a background check, but it wasn't comprehensive. Annie's job was analyzing drug samples confiscated by cops. Cops can't come to court and say it's cocaine. A cop can say I thought it was cocaine, it was packaged like cocaine, and it looked like cocaine. You need a chemist to come in, and say: "I performed a molecular analysis. This drug was cocaine." So between 2002 and 2012, Annie Dookhan analyzed sixty-thousand drug samples and gave testimony on thirty-four thousand in court or via sworn affidavit. In 2012, she was caught. She wasn't analyzing anything. Whatever the cop said it was, she signed off on it. Cop said cocaine, heroin, or meth so she said cocaine, heroin, or meth. "Gordon, most drug labs have a secondary sampling process where they check results a second time to make sure the first findings were correct." Yes, and the drug abuse lab had such a process. The secondary sampling process was properly designed, and it was up to date. The problem was nobody followed it. Annie Dookhan was getting away with her behavior because cops know their drugs. Secondary sampling process would come up with the same findings.

STRATEGIC HINT FOR YOUR CONSIDERATION

The best predictor of future behavior is past behavior, and if you hire people who have a history of flawed behavior or deficient integrity, then you have a problem lying in wait.

Every now and then, the secondary sampling chemists would arrive at a different result than Annie Dookhan. That's a big deal, and they had a process on what to do in such situations. That process was also properly designed and up to date. Nobody, however, followed the policy, which was clear: If the second chemist derived a different result, then a supervisor shall be notified. If you notify the supervisor, however, then Annie will get in trouble, and she's a delightful single mom with a special needs child. So the secondary chemists would talk to her rather than report the error. The deviation became the norm. She continued to misbehave until she finally was caught several years later.

If I'm a public defender, do I have a good argument that my client needs to be released from prison? Yes, and public defenders have asked for and received millions of dollars to release the allegedly wrongfully convicted from prison. Conversely, the attorney general has asked for and received millions of dollars to keep these people in prison. All could've been prevented if the background investigation performed on Ms. Dookhan was comprehensive. There were inaccuracies in her application, one of which was a Master of Science degree from the University of Massachusetts. No one chose to confirm the accuracy of that statement, and it turned out to be false. Bottom line: Ms. Dookhan was improperly hired, and her conduct was exacerbated by a failure of supervision. This combination of a cursory background investigation and improper supervision cost the taxpayers of Massachusetts one-billion dollars. All could've been avoided had the State of Massachusetts properly followed their system for onboarding new employees. And that correlates to the importance of the four interrelated factors of getting and keeping good people: recruitment, background investigations, probation periods, and performance evaluations.

STRATEGIC HINT FOR YOUR CONSIDERATION

Quality and compliant information is crucial for hiring the right candidates, but there's no perfect background check or single database that includes all civil and criminal records.

The Annie Dookhan example crystallizes the importance of comprehensive background investigations that are thorough, deliberative, and unrushed. Regrettably, there are many Annie Dookhans lying in wait within all types of organizations. That's why you must have rules and systems for this high-risk, discretionary time task. I've created ten such rules for the improvement of background investigations, which I call GRIBI© (Graham's rules for improving background investigations). These rules encompass the core areas an executive, manager, and supervisor should know regarding this essential process. Its purpose is to maximize employee screening and prevent consequences (tragedies) by filtering out bad actors. Comprehensive background investigations are one of the four interrelated steps in augmenting pillar one: getting and keeping good people. Figure 5.2 illustrates these ten sequential rules and provides the platform for further examination.

GRIBI© #1

Understand the Purpose and Importance of Good Background Investigations

Background investigations are one of the most important activities performed by water-related entities. The women and men you select today will interact with our great public tomorrow, and they'll be the leaders of your organization in the next

Graham's Rules for Improving Background Investigations (GRIBI©)		
GRIBI© #1	➲	Understand the purpose and importance of good background investigations.
GRIBI© #2	➲	Establish a policy (system) regarding background investigations.
GRIBI© #3	➲	Select people who have the knowledge, skill, ability, and desire to conduct background investigations.
GRIBI© #4	➲	The process shouldn't be a secret.
GRIBI© #5	➲	Take the time to do the job right.
GRIBI© #6	➲	Utilize all available resources, and don't worry about the cost.
GRIBI© #7	➲	Remember the importance of accuracy.
GRIBI© #8	➲	Always proofread your documentation.
GRIBI© #9	➲	Supervisory review is essential.
GRIBI© #10	➲	Learn from your experiences.

FIGURE 5.2 Graham's rules for improving background investigations (GRIBI©). Source: Graham Research Consultants, LLC.

generation. It's imperative you have a process that screens out those who are unwilling or unable to do the job they've been assigned. Background investigations provide that process because the best predictor of future behavior is past behavior. Past habits are indicative of future habits. The purpose of a background investigation is to identify and document relevant past behaviors so you have sufficient information to make an informed decision. Everyone has the right to seek a job in your organization, but you have an affirmative and ethical obligation to select individuals who have the desire and attributes to do the assigned job. The jobs within your profession are complex and critical to a functioning society, and, therefore, require special people. This complexity includes executives, managers, supervisors, and line-personnel. Not everyone is qualified to be employed in your organization.

STRATEGIC HINT FOR YOUR CONSIDERATION

The development of waivers, written by competent counsel, specific to the job description you're testing for and specific to the applicant is necessary to discourage industry recycling of bad actors.

With this in mind, the following is essential: If there's a dispute between the rights of the applicant and the rights of the public, always make your decision in favor of the public. Public trust and confidence are essential for organizational survival. The core ingredient of a successful water/wastewater professional is integrity. Absent integrity, you don't have anything. A single employee demonstrating a lack of accountability and integrity can negatively impact the entire profession. Please take that assertion seriously, and remember an applicant puddle is a problem lying in wait. You must revisit the recruitment process. It's also important to recognize the importance of diversity. If your workforce doesn't look like your community, then you have another problem lying in wait. One final thought that I've stressed in many presentations to public safety professionals, but also applies to water/wastewater professionals, is I recommend there be an ongoing background investigation process throughout your organization. The federal law enforcement model of every five years is a good idea. Some of your good people go bad over time, and the only way you learn of their conversion is post-incident. You must have a process to ensure every employee continues to possess the required traits of accountability and integrity at point of hire. There's no other way to prevent this scenario than ongoing background investigations.

GRIBI© #2

Establish a Policy (System) Regarding Background Investigations

As with any other high-risk activities, your organization needs a policy (system) regarding background investigations. The policy needs to be specific to your organization, and because of the high-risk nature, it needs to be reviewed by competent counsel on a regular basis, including after its development. Depending on your jurisdiction, your policy may need to include criteria for the selection of investigators, information sources, access of medical and other confidential information, retention, redaction and release of information, and other relevant considerations. Of importance is the development of waivers specific to the job description you're testing for and specific to the applicant.

STRATEGIC HINT FOR YOUR CONSIDERATION

Develop a long-term relationship with your backgrounds vendor, verify what they're doing for you, and periodically review the type of reports they're pulling so you can evaluate their appropriateness for your organization.

As mentioned earlier, there are many laws restricting what information can be gathered, and many records are confidential in nature. Appropriate waivers are essential, and these need to be drafted by your competent counsel. There are many pitfalls in the genre of employment law, and these laws vary from jurisdiction to jurisdiction. The purpose of any policy is to set standards so as to maintain a consistent approach between multiple people who are doing the particular activity. If you don't have any policy now, you need one. Don't reinvent the process. Most large water-related entities have a policy on backgrounds. Another suggestion is to contact your state police and request a copy of their policy. You may not require all their investigative background tools, but it'll be an excellent encapsulation of a thorough policy. Don't swipe another organization's policy without first assuring it's right for your organization. Otherwise, you'll have a problem lying in wait that can end with disastrous consequences (tragedies). What's good for one organization isn't necessarily good for you. Prudent investment in pre-incident prevention is a necessity to avoid an Annie Dookhan situation and the resulting organizational embarrassment, civil liability, and loss of community goodwill.

GRIBI© #3

Select People Who Have the Knowledge, Skill, Ability, and Desire to Conduct Background Investigations

Not everyone in your organization is qualified to be a background investigator. The job is tedious, requires a lot of documentation and meticulous review, and is intensive in areas that aren't all that exciting. However, it's critical to the survival of your organization. Many organizations aren't continually hiring people, so background investigations are, at best, a part-time occupation and fall into the other duties as assigned category. If the preceding has to be, then please ensure the person doing this high-risk activity has the necessary time and competencies to properly do it. Only qualified and proven investigators with training specific to background investigations should perform them. If you don't have such an investigator, water-related entities in a geographic area can pool resources to fund this person and process. Also, there's a learning curve involved in becoming a proficient background investigator. So if you're choosing people to do this job who are near retirement, it might be a mistake unless they're going to continue to do the job as a member of a Deferred Retirement Option Plan (DROP) or as a retired annuitant. It takes about two years of solid work to pick up the necessary skills to do this job correctly.

Please don't make the background investigator position one where you rotate people in-and-out annually. That's not good. I've also heard stories about background investigations being a depository for people who failed at every other job in the organization. I wonder how that'll work out in the long run. Additionally, there's a temptation among some organizations to contract out the process to a private investigation firm. This too, can be dangerous. Investigate the investigator up front by asking for references and past performance. Develop a long-term relationship with this firm and verify what they're doing for you. The vast majority of outside investigators are dedicated, hard-working people, but there's always that small percent who are simply in it for the money. With that in mind, recognize the pitfalls of flat-rate payment per investigation. I know this viewpoint isn't popular, but I've witnessed situations where private investigators run into complexities in a background investigation and don't follow through to properly finalize their findings. If you must contract this function, then pay a fair rate by the hour, not by the investigation. And have some processes (audits) in place to confirm the findings of the outside investigator. Or perhaps you could use retired personnel to assist in this process. They still have substantial ties to your organization, and it'll free up your personnel to perform their primary job.

STRATEGIC HINT FOR YOUR CONSIDERATION

Allowing the public to see the details in selecting quality personnel will benefit your goodwill, and disclosing the process to potential applicants will discourage those who know their past behaviors will be discovered from applying.

GRIBI© #4

The Process (System) Shouldn't Be a Secret

While there are some portions of your policies (systems) that need to remain confidential, the background investigation process should be available for inspection by the public and by applicants seeking employment. Allowing the public to see the details and your efforts in selecting quality personnel will benefit your community goodwill. Additionally, disclosing the process to potential applicants will likely discourage those who know their past behaviors will be discovered. Once they realize how detailed your investigation will be, they may opt out early. That'll save your time and theirs and will prevent many subsequent problems from occurring. Additionally, it will reduce organizational expenditures in this important activity. A specific waiver for lateral hires will similarly save you grief by deterring bad actors currently employed by other water-related entities from making applications to your organization and tying up your precious background investigation resources.

The public must know there's a process (system) to maximize honesty, integrity, and competence of the people seeking employment with your organization. You and your family are members of the public, so please take it seriously. Secret policies are suspicious, and they lead to distrust among line-personnel and organizational executives, managers, and supervisors.

STRATEGIC HINT FOR YOUR CONSIDERATION

Background investigations and recruitment are separate units with different responsibilities, so they need to be separated in your organization. The former is internal affairs, whereas the latter is public affairs.

GRIBI© #5

Take the Time to Do the Job Right

Background investigations are a discretionary time activity. Your investigators should have the necessary time to gather all the facts about an applicant's past behaviors and present situation. Rushing a background investigation is like hurrying up a felony stop or a door kick for cops and a line break for water/wastewater professionals. Nothing but bad will come from rushing this activity.

Use discretionary time to gather all the relevant and legally obtainable facts from an applicant's past. You won't be able to do the background correctly if you're rushed. Here's an example from a public safety organization (in a city where I used to live that's now home to the Queen Mary) to accentuate the point. This organization found some extra money one year but needed to spend it by year end. So you know the rest of the story. They hired twenty people and rushed to fill an Academy class. This approach is wrongheaded. Depending on the workload of the investigator, it may take from six to eight weeks to perform a good background investigation, and this timeline may be extended if the applicant has been out of country or has some personal issues that need more attention.

Please take the time to do the investigation right. Use your discretionary time to gather all relevant and legally obtainable facts from an applicant's past. Set up the investigative file in a systematic manner so information is processed and logged quickly and accurately. Respond to leads and follow them as necessary. Such a methodical system will allow you to see at a glance if an applicant or other source hasn't responded to your queries. Don't prematurely close a file assuming the remaining information is non-essential. Trust me on this one, and it bears repeating. Don't prematurely close a file. Wait and gather all the information prior to forwarding it for a decision. Background investigations and recruitment are separate units with different responsibilities, so they need to be separated in your organization. Recruitment of new personnel is a responsibility of public affairs while background investigations are a function of internal affairs.

GRIBI© #6

Utilize All Available Resources, and Don't Worry about the Cost

STRATEGIC HINT FOR YOUR CONSIDERATION

Contact the references on the personal history statement (PHS), and ask each reference for the name of another person that you can contact to ascertain applicant suitability for the job.

There are many available resources to verify information offered by the applicant and to find new investigative leads. Examples include: interviews with the applicant; listed reference checks; secondary sources discovered after these reference checks; interviews of acquaintances, former spouses, and landlords; residence checks; education checks; applications for other organizations; employment checks; military service checks; legal and criminal history checks; motor vehicle checks; and financial checks. Use the stated references on the personal history statement (PHS) and contact them. Obviously, anyone the applicant lists is likely to provide a favorable reference. Be sure to inquire of each reference person the name of another person that you can contact to ascertain suitability for the job. There are massive databases maintained by law enforcement agencies that are accessible subject to the applicable laws of your jurisdiction. Numerous private sector companies have similarly procured public records from around the United States and store them in databases for purchase by interested parties. Caution: Only use this information as an investigative lead. Don't base your decision or opinion of an applicant solely on this information. Some of it is stale, wrong, or filed under a similar name.

Compare and contrast what you learn through these databases with what's stated as factual on the personal history statement (PHS). If there are inconsistencies, check those out and make inquiries of the candidate. Since it involves rights of privacy, check with your competent legal counsel prior to accessing these databases. Take the time to do this activity right. What's the line I use as a lawyer? You can pay me now, or you can pay me a lot more later. This applies to everything we do: Prevention is better than correction. In too many aspects of our society, we're light on prevention because it requires a certain expenditure of money to address an uncertainty in the future. Please remember that if an employee is ever involved in something terrible, the downstream lawyers will perform their own background investigation of the accused. They'll spend the necessary resources to ascertain all available facts of the employee prior to being hired by your organization. Don't be parsimonious, as nothing but bad will come from such an action.

GRIBI© #7

Remember the Importance of Accuracy

Once the package has been submitted and the decision is made, you're wedded to the package and decision forever. Remember, facts are facts and opinions are opinions. It's critical that each fact be accurately recorded. In any downstream litigation, whether it be external (e.g., negligent hiring) or internal (e.g., wrongful termination), your investigative package is subject to discovery and can't be changed from what you originally submitted. Full and accurate documentation of each step of the process is essential.

STRATEGIC HINT FOR YOUR CONSIDERATION

In any downstream litigation, whether it be external (e.g., negligent hiring) or internal (e.g., wrongful termination), your investigative package is subject to discovery and can't be changed from what you originally submitted.

GRIBI© #8

Always Proofread Your Documentation

Proofreading will reinforce the importance of accuracy and assure conformance with rule number seven. Small things are important in the long run so spend time reviewing what you've documented prior to submitting it. Pretend you're an attorney suing your organization for failure to do the background investigation properly. You discovered a copy of the background investigation policy and hired an expert witness to review the investigative file while comparing it to the policy. If you're comfortable with this scenario, then you've probably performed the activity correctly. Be careful in summarizing material you've received. Is it a fair and impartial summary containing the good and the bad? Finally, your review is far more than protection from liability. If you don't perform the job right, you may be hiring someone who'll be a problem lying in wait. Or, your failure may deny a qualified applicant the opportunity to work for your organization. In either case, there are only losers in the equation. Do the job right, document it right, and proofread the entire file.

GRIBI© #9

Supervisory Review Is Essential

The background investigation process is high risk, and there are downstream consequences for not performing the activity correctly. Minimize risk and exposure by incorporating knowledgeable supervisors into the review process and not just at the terminus of investigation. Your investigators will rise and fall to your level of expectation. Set high standards, and that's what you'll get. Set no, low, or inconsistent standards and that's what you'll get. The integrity of the process is essential, and achieving good results is never an accident. Finally, if we want a thorough and proper review of a background investigation, the reviewer must understand what she's reviewing. It's essential the supervisor has experience and training on the activity.

GRIBI© #10

Learn from Your Experiences

STRATEGIC HINT FOR YOUR CONSIDERATION

You should perform periodic background investigations to make sure the good person you hired is still the same good person.

STRATEGIC HINT FOR YOUR CONSIDERATION

If you don't perform the background investigation correctly, you may be hiring someone who'll be a problem lying in wait. Or, your failure may deny a qualified applicant the opportunity to work for your organization.

As with any activity, you need practice to become proficient. As you learn, share your knowledge with your peers. If your newfound knowledge isn't in the policy manual, then make appropriate revisions to update the policy on this better way. Joining a professional group of similar investigators (like the California Background Investigators Association) will benefit you and your organization. There's a wealth of knowledge to access. Finally, background investigations are the only tool available to gather the necessary facts about applicants so take the process seriously, and everyone will benefit. Ethical behavior in your organization starts by hiring people who possess integrity.

FINAL COMMENTS

Throughout this chapter, I've mentioned the importance of initial background investigations. Initial (prior to hire) background investigations are absolutely essential, but I recommend the process be extended into the future. Every five years, you should perform another background investigation to make sure the good person you hired is still the same good person. In many instances, the involved bad actors went through a background investigation at the start of their career. Over time, they began behaving inappropriately and exhibiting traits that would've rejected them as an applicant. Since these evolved bad actors were already employed and knew there wouldn't be another background investigation, they had less fear of being discovered. Your control measure (system) for this scenario is an ongoing background investigation process. I recognize there's a certain cost today to address potential future problems, but I'm getting more support on this issue. A parallel for water-related entities would be public safety organizations. Look at how many cops are rejected when the sheriff's office annexes a smaller police department and performs background investigations on the current cops in that department.

Returning to the people pillar, so far we've reviewed the importance of recruitment efforts and comprehensive background checks. Let's look at some additional aspects. Understand the probationary process: Your trainers need to take this process seriously, and so do your supervisors. If they have people on probation who either can't or won't do the job, you must act immediately. They won't get better over time. Such actions correlate to the role of supervision in your organization. Our next chapter will detail the importance of the supervisory position and the skills necessary to ascend to a high-reliability organization. These attributes, which we'll examine, are compulsory if you want your employees to do the right thing, the right way every time. We know the outcome of such behavior is an enterprise underrepresented in tragedies and overrepresented in employee fitness and customer loyalty. The role of your supervisors is the ballast of organizational accountability and integrity. Next stop is maximizing supervisory performance. Until then, work safely.

Chapter Takeaway on Media Relations/Crisis Communication

Please refer to the media relations/crisis communication material in the Addendum section and ask yourself how you would address the following scenario:

You just returned from an afternoon meeting with your general counsel and county district attorney. How could Jim, your controller of twenty years, steal nearly five-million dollars from your organization? Worse, how did you not find out until seven years later? Your insurance policy only has a one-million-dollar limit for employee dishonesty claims. That means your ratepayers will have to pay the rest, and you already increased their monthly rates by twenty-seven percent to address deficient reserving practices. This tragedy will hit the local news tonight and the morning papers tomorrow. Your organization has audited financials and hires an outside auditing firm to perform integrity tests. For the past two years, the auditor inquired about the strength of your internal controls. She expressed concern that they weren't as robust as your peers, and she reminded you her job is to express an opinion on your financial statements and to ensure documents are free from material misstatement. And auditor's job isn't to express an opinion on the effectiveness of organizational controls.

The auditor's concerns were shared with your finance director, but she argued the cost to implement them would be high and unnecessary because she and Jim started with the organization the same time many years ago. There was nothing to worry about. It gets worse. Your general counsel opened an investigation and performed an exhaustive background check on Jim. You compared her report with the original, and her findings were shocking. The original report was cursory and basic. It centered on criminal history and open source data. Your general counsel hired a professional investigator who confirmed Jim was fired from a previous job and removed from a local nonprofit for alleged but unproven financial irregularities. It was also discovered that Jim never received his accounting degree like he disclosed. This information was not verified on the original report. And the worst part was a recent credit check showed massive debts going back fifteen years to gaming organizations across the country. An ongoing background investigation report would've spotted that red flag.

It's clear this financial crime was preventable. The impact to your organization's credibility and goodwill will be severely tarnished. The ratepayers, who already objected to your recent rate increase, will not be sympathetic to the situation. You must address the situation tomorrow morning through a press conference. How do you address the issue with candor and anger without acknowledging all the internal blunders? And how can you come across as authentic in light of these errors? Finally, you know the press will cast you as an incompetent leader overseeing an organization with a purported history of bad practices. Do you address that concern as well? You have an evening to collect your thoughts before the barrage of questions and scrutiny. How will you prepare?

6 Meaningful Performance Evaluations

Do Them Right or Get Rid of Them

2019© Photo Courtesy of Placer County Water Agency, Placer County California 95604

STRATEGIC HINT FOR YOUR CONSIDERATION

Performance evaluations are a great risk management tool when taken seriously and a reverberating cudgel when the system isn't properly designed or implemented.

DOI: 10.1201/9781003229087-6

SUMMARY

Performance evaluations are a meaningful risk management tool when taken seriously and a problem lying in wait when not. Properly deployed, these documents improve employee performance by assessing progress and addressing areas of necessary development.

The five pillars of success apply to performance evaluations. If you want them done right, you need good people preparing them, a validated policy to show good people how to do them, training to ensure good people know how to do them, supervision actively involved in the process, and discipline when the process isn't taken seriously.

Chapter emphasis will center on ten specific rules for preventing tragedies from inappropriately prepared and improperly documented performance evaluations. These rules will ensure this high-risk, essential activity retains its organizational value.

Hello again. Gordon Graham here, and welcome back to our journey toward *real* risk management. Chapter 6 will continue our granular examination of organizational risk management, which is where theory is applied in practice. High-reliability organizations address their problems lying in wait by probing deeper than proximate cause and carefully investigating the root cause of tragedies. This methodical analysis prevents similar tragedies from reoccurring by integrating the root causational factors involved in tragedies (people, policy, training, supervision, and discipline) into proactive systems. By converting their problem factors to solution factors, these enterprises achieve the five pillars of success and ascend to a high-reliability organization.

The five pillars of success is concisely described as follows: Good people who act with good policy, who are regularly trained, properly supervised, and in an organization that addresses arrogance, ignorance, and complacency with fair and impartial discipline will prevent tragedies and maximize customer loyalty by doing the right thing, the right way every time. We began our examination of pillar one, getting and keeping good people, in Chapter 5 by reviewing the importance and purpose of comprehensive background checks. This component, when combined with recruitment, probation, and performance evaluations, maximizes the likelihood of success for pillar one (people). We'll continue to delve into the people pillar in this chapter, "Meaningful Performance Evaluations," by examining the right way and wrong way to perform this high-risk activity.

Performance evaluations are a great risk management tool when taken seriously. The same tool, however, can reverberate in ways you've never envisioned when the evaluation system isn't properly designed or implemented. This chapter will impart in depth thoughts on the importance of performance evaluations and proven strategies to improve the quality of the process. Figure 6.1 illustrates the interrelated

Getting & Keeping Good People

Four Interconnected Factors

- Recruitment
- Background Investigations
- Probation
- Performance Evaluations

FIGURE 6.1 Getting and keeping good people. Source: Graham Research Consultants, LLC.

components of recruitment, backgrounds, probation, and performance evaluations in transforming the problem factor of people into a solution factor of getting and keeping good people for your organization. High-reliability organizations deploy performance evaluations through accurate employee appraisals and clarity of process. This tool, when adeptly handled, is an embodiment of *real* risk management.

I have a different view of performance evaluations than most people, and I would like you to reconsider your view of the entire process. I'll start with this thought: Do away with performance evaluations. As a lawyer, I've learned performance evaluations are written documents prepared annually (allegedly) without a lot of thought and then lie in wait until they come back and boomerang the organization, which happens on a regular basis. I've learned to loathe performance evaluations because they pose risk, and lawyers have a bias for risk aversion. I also wear my risk management hat. As a risk manager, I embrace performance evaluations with one caveat: You must take them seriously. A properly prepared performance evaluation is an excellent risk management tool. It's a regular opportunity to assess how a given employee is doing and provides appropriate control measures (systems) to address future risks. Such documents have the goal of improving employee performance. But here's the rest of the story: I'm not aware of any water-related entity that takes performance evaluations seriously. The end product is unimpressive, and everyone knows they're a problem lying in wait. Deep down, I know you agree with me. With the advent of word processing some thirty years ago, performance evaluations have turned into a search/replace exercise in which a supervisor pulls up the last evaluation prepared for a given employee and asks herself: "How much do I have to change to make it look fresh?" In some cases, the supervisor may use paragraphs of the same text for different employees. Word processing allows this shortcut; poor management permits it to continue.

STRATEGIC HINT FOR YOUR CONSIDERATION

A proper performance evaluation system includes meaningful job descriptions, identified objectives, validation process for objectives, employee performance repository, future goals, and validated rating system.

If you actually believe your organization is taking performance evaluations seriously, then I have a challenge for you. Assuming you have the authority, go to your organizational personnel files and pull ten performance evaluations at random that were prepared last year. Then, go back one more year and pull the same employee evaluations and compare the evaluations. I guarantee at least one of the ten sets (possibly more) will be identical, with the exception of the date. If you want to continue this process, take those ten employee names to your human resources personnel and ask: "Have you had a negative contact with any of these ten employees over this two-year window?" At least one of the employees will have had a negative contact. And what's said about that negative contact in the performance evaluation? Nothing. If you really want to continue, take the supervisors of these ten employees drinking. Get them up to the level where loose lips sink ships and ask: "So what do you really think about these employees?" I guarantee you that what they'll tell you is more accurate than what they wrote in the performance evaluation.

Here's the bottom line: Too often, your water/wastewater professionals are overrated every year. To be fair, there's occasionally a supervisor who for some reason doesn't like an employee, and they'll deliberately underrate that person. That's extremely rare, and when it occurs, the given employee will usually raise objections and there will be some type of investigation to determine the validity of the poor performance evaluation.

Why does this scenario repeat itself? One reason is bias. Supervisors and managers generally like their people. They've worked together, they know their spouses and kids, and their employees are generally good people. Even if the supervisor wants to be accurate, there's a built-in bias in favor of their personnel. The real reason water/wastewater professionals are overrated is it's easy. No one complains when they get overrated. If someone thinks they've been underrated, they'll object to organizational management, and there will be an investigation to determine the accuracy of the evaluation. The supervisor will be scrutinized: "I hope you can justify the ratings, Nancy." It's much easier to overrate people because there will never be a complaint. That's exactly the situation with water-related entities. If you take people on and try to improve their performance, and they object, you'll likely be facing an allegation of bad behavior yourself. Ultimately, you get paid X. If you overrate people to avoid problems, you still get paid X. When mediocrity or poor performance is rewarded, it spreads. Here's a hint: Your personnel share their performance evaluations. When a known indolent receives evaluations that exaggerate his organizational usefulness, you've just told every other employee there's no need to work hard, for they, too, will be equally rewarded for doing nothing. The system is self-reinforcing. Even new supervisors who want to do the right thing will be reprimanded if they attempt to give honest performance evaluations. Worse, these supervisors can be accused of harassment, bias, or discrimination.

STRATEGIC HINT FOR YOUR CONSIDERATION

A validated rating system only needs three categories: meets standards; exceeds standards; and doesn't meet standards. Either people are doing their job or not, and there's nothing wrong with receiving a "meets standards" evaluation.

Assuming your current performance evaluation system is properly designed, you may simply need to implement some internal control measures (systems) to warrant the process is taken seriously. What constitutes a properly designed performance evaluation system? It's one that has the following components: meaningful job descriptions; identified objectives for each job; a process to ensure employees are meeting these objectives; a method to collect and analyze employee performance data; goals for the next reporting period; and a validated rating system. With respect to this rating system, you need three categories: meets standards; exceeds standards; and doesn't meet standards. Over the years, I've seen up to nine categories of performance. That's unnecessary. Either people are doing their job or not. Some (the ten percenters) will exceed standards and some (the other ten percenters) will not. Everyone else will be in between.

There's nothing wrong with receiving a "meets standards" evaluation. For those not meeting standards, you must deploy a performance improvement plan to augment their development and monitor their progress. If the employee can't or won't meet standards, it's time to separate them from the organization. You'll also need strict management control of the performance evaluation process. All performance evaluations should be reviewed by management prior to being signed by the employee and supervisor. You similarly need a robust audit process to ensure performance evaluations are taken seriously by supervisors writing them and their managers reviewing them.

STRATEGIC HINT FOR YOUR CONSIDERATION

During the annual performance evaluation process, review the harassment policy with every employee so there's documentation of knowledge and of available remedies should there be a problem.

If you're committed to your current system, it'll be difficult to implement these changes instantly. How do you transition from a system in which everyone is overrated to one where people are honestly evaluated? You can take a moratorium from the original process, inform employees that there will be a commitment to a system in which people are evaluated honestly, and then gradually move to the new system. Here are a couple of ideas to improve the process. You should integrate an annual test that focuses on each employee's core critical tasks. Every water/wastewater professional has a small number of activities that carry a real risk of tragedy if they aren't performed correctly. These core critical tasks must be identified through proper risk assessments and risk/frequency matrixes, and your personnel must be tested on these activities regularly. One of my irritants is that after you're released from probation (assuming you choose not to promote), you've taken your last serious test. The only time we realize an employee doesn't know how to perform a core critical task is after the activity ends badly.

The same applies to harassment policies. The only time we find out a water/wastewater professional doesn't know the harassment policy is after a lawsuit is filed. It's then too late, and you're in Lawyerville. Moreover, an annual test on core critical tasks and critical tasks would help prevent tragedies. The only acceptable pre-incident knowledge level verification would be one-hundred percent. When all of your people are achieving this score, make the test tougher. Our public and personnel deserve better than minimum standards. "You'll never get this testing idea past the union!" Why don't you ask the union to help write the test? "If we have the union help, the employees will know the answers!" And your point is?

The most expensive lawsuits facing water-related entities are human resources (HR) related; specifically, allegations of a hostile work environment with regard to protected class status. As a lawyer, I must prove five things if I'm suing your organization for a hostile work environment: (1) My client is a member of a protected class; (2) there are behaviors in the workplace that negatively impact my client; (3) the organization knew or should've known about these behaviors; (4) the organization failed to stop these behaviors; and (5) due to your failure to act, my client has suffered a harm or loss and is owed damages. In lawyer-speak, that's the burden of proof. In risk management terms, that's an identifiable risk and a manageable risk. Enter a tool I call APE (annual performance evaluation). I designed this document in the 1980s and it's still valid today. Many private and public organizations have used it with success. Here's the purpose of APE: During the annual performance evaluation process, the supervisor (or HR personnel) is required to review the harassment policy with every employee. The employee answers some questions (documented on the evaluation) regarding their knowledge of the policy and what to do if they ever have a problem.

STRATEGIC HINT FOR YOUR CONSIDERATION

Performance evaluations are a useful management tool that can improve morale of employees who meet expectations and provide fair warning to employees whose performance is unsatisfactory and requiring improvement.

Of what value is APE? When an employee sues your organization for hostile work environment, and they list all the allegations of systemic bad behavior over several years and claim they didn't know what to do, this document might save a year in expensive lawyer fees. Make sure your competent counsel reviews any changes to your employee evaluation process. Regardless of your process, you must make a commitment to gathering data (to support the evaluation). Make it your goal to document each employee's performance every week. At the end of the rating period, you'll have plenty of specific, articulable facts (SAF) to support your ratings. If you perform this activity diligently, there will be no surprises for the given employee, and the time you spend preparing the final document will be reduced tremendously. That leads to the final mistake water-related entities make with personnel reviews: allowing them to be fodder for retaliation suits.

Let's say an employee files a hostile work environment case against your organization. These cases are difficult for employees to win, but employment lawyers know a little patience can turn their case into a retaliation suit. Not following? Year after year, the supervisor provided said employee a good evaluation because that's the organizational culture. When the supervisor is named as a defendant in the hostile work environment case, he's now angry and may honestly rate the employee as needs improvement. To a future jury, this honest evaluation is evidence of retaliation against the employee for filing the hostile work environment claim. Such a scenario may seem far-fetched to you, but in every employment law case I handle, performance evaluations come back to afflict the involved organization because indifferent supervisors continue to overrate employees.

Performance evaluations require careful, consistent, and deliberative handling. It's paramount your organization develops a performance evaluation system that's properly designed, kept up to date, and fully implemented. Allow me to introduce GRIPE© (Graham's rules for improving performance evaluations), which are my rules for the improvement of performance evaluations. This system comprises ten activity-specific techniques for maximizing the utility of these documents while preventing the consequences caused by inappropriate preparation or improper documentation. GRIPE© plays an integral role in pillar one: getting and keeping good people. These rules are summarized in Figure 6.2. Let's examine each one in greater detail.

GRIPE© #1

Understand the Purpose of Employee Evaluations

Performance evaluations are a useful management tool that can improve morale of employees who meet performance expectations and provide fair warning to employees whose performance is unsatisfactory and where improvement is needed to meet standards. Performance evaluations are critical to a well-run organization. Only a few water-related entities are subject to laws that require performance evaluations for their employees. Like the private sector, performance evaluations in most public-sector organizations are optional. With this in

Graham's Rules for Improving Performance Evaluations (GRIPE©)

GRIPE© #1	➲	Understand the purpose of employee evaluations.
GRIPE© #2	➲	Get a policy (system) regarding evaluations.
GRIPE© #3	➲	Select people to be supervisors who are capable of being honest with employees.
GRIPE© #4	➲	The process shouldn't be a secret.
GRIPE© #5	➲	Before you put pencil to paper (or fingers on a keyboard), think.
GRIPE© #6	➲	Be specific during the evaluation process.
GRIPE© #7	➲	Remember the importance of accuracy.
GRIPE© #8	➲	Always proofread your documentation.
GRIPE© #9	➲	If you're right, don't change it.
GRIPE© #10	➲	Learn from and share your experiences.

FIGURE 6.2 Graham's rules for improving performance evaluations (GRIPE©). Source: Graham Research Consultants, LLC.

mind, remember performance evaluations are an excellent management tool if your organization is committed to completing them correctly. Otherwise, you're better off without them.

Ill-prepared performance evaluations are worse than no performance evaluations. Poorly prepared or inaccurate evaluations can lead to problems downstream, including discipline hearings, arbitration, grievance procedures, and litigation. Preparation of performance evaluations is high risk, and like any other high-risk activity, it requires special consideration. The five pillars of success apply to performance evaluations: If you want them done right, you need good people preparing them, a validated policy to show good people how to do them, training (initial and ongoing) to ensure good people know how to do them, supervisors (this also means executive and management oversight) actively involving in the process, and disciplining when the process isn't taken seriously.

GRIPE© #2

Implement a Policy (System) Regarding Evaluations

As with any other high-risk activity, your organization needs a policy (system) describing how to perform this process. The policy must be organization-specific and reviewed by your competent counsel on a regular basis. The purpose of any policy is to set standards and maintain a consistent approach toward performance. Your organizational policy in this regard must be specific to each job description. Well-prepared job descriptions and performance expectations are essential precursors to the evaluation process. These instruments must be codified before any attempt can be made to evaluate performance. Observations by supervisors and managers of employee performance throughout the rating period must be collected and saved via appropriate documentation.

STRATEGIC HINT FOR YOUR CONSIDERATION

The performance evaluation process should include an assessment of knowledge regarding the involved job description so training can be delivered to specific areas where there's a demonstrated need for improvement.

Your policy should include direction on how to analyze data to establish a rating, and, as much as possible, this portion of the process must be highly objective. The policy should similarly include direction on how to gather necessary information from the rated employee and other sources, how to prepare the document, and how to finalize the document. Management review of the prepared document is essential, preferably before the document is reviewed and signed by the rated employee. This management review should include auditing the prepared document for accuracy, meeting with the rated employee, discussing their performance, looking for potential problem areas, and pre-incident verification of knowledge level for high-risk, low-frequency activities within their job description. It's also an excellent time to check for licensing validity, other certifications, and up-to-date résumés. I'm an advocate of continuous improvement involving quantifiable metrics, so collaborate with the employee in finding better ways to reach that goal. Finally, the performance evaluation process should include an assessment of knowledge regarding the involved job description so training can be delivered to specific areas where there's a demonstrated need for improvement.

GRIPE© #3

Select People to Be Supervisors Who Are Capable of Being Honest with Employees

STRATEGIC HINT FOR YOUR CONSIDERATION

Thorough and advance review of performance evaluations by informed managers can prevent intentional or negligent overrating by supervisors who either lack the fortitude to be honest or want to be friends with everybody.

STRATEGIC HINT FOR YOUR CONSIDERATION

Keeping the performance evaluation process secret will lead to less-than-optimum performance and low morale. When employees know each step of the process, they're better motivated to meet the standards you've set.

Some performance evaluations are properly prepared, but many exaggerate employee performance. We like the people we work with and work around. This bias can pose a major problem during the performance evaluation process. Some supervisors use the process inappropriately to get even with people they don't like, but more frequently, the performance evaluation overrates employees. Overrating occurs in two ways: intentionally and negligently. Both can be prevented through a review of the process by informed managers. Most employees are overrated because supervisors lack the fortitude to be honest with the person they're rating. Some supervisors want to be friends with everybody, some believe rating people high will improve their performance, some believe they have to rate people as they've been rated. Each of these reasons leads to organizational grief (tragedies). When you're promoting people in your organization, make sure you evaluate their future ability to be honest with employees, including during the rating process. Finally, ensure you have a training program for all supervisors and managers so everyone understands what evaluations are about, the consequences of doing them wrong, and most importantly, how to do them properly. It's the right thing to do.

GRIPE© #4

The Process (System) Shouldn't Be a Secret

After deriving a good policy (system) and ensuring you've good people doing the rating and reviewing of the evaluations, make sure all employees are aware of how the process works, including how data is collected and analyzed for use in the evaluation. Good people (and I assume you have them in your organization) will rise and fall to your level of expectation. The evaluation process must be explained in your employee guide so everyone knows how their performance will be evaluated. Keeping the process secret (or creating the perception that it's secret) will only cause future problems including less-than-optimum performance and low morale. When employees know each step of the process, they're better motivated to meet the standards you've set.

GRIPE© #5

Before You Put Pencil to Paper (or Fingers on a Keyboard), Think

As mentioned above, performance evaluations are a high-risk activity, but they're also discretionary time in nature. That means you have a plenty of time to think before you perform the activity. Use this time to review the policy of your organization and make sure you understand what you're doing. Develop an action plan as to how you're going to prepare this document. Utilize your discretionary time to read the last evaluation of the involved employee. If she had problem areas, or was given direction on areas needing improvement, check to see if she followed the recommended direction. Finally, use your discretionary time to analyze her performance during the rating period and talk to other supervisors for their input regarding her performance.

This discretionary time can be used to have a pre-evaluation discussion with the involved employee to discuss what you're doing, to share your preliminary data analysis, and to enlist her feedback and input regarding her specific performance. After you've done that and prepped a rough draft, you can use this discretionary time to check with your manager or human resources (HR) personnel to get their input into your efforts. This review by a manager or HR personnel may prevent some of the common errors made in the evaluation process. Remember you're the one preparing the document, so weigh the comments of others with your knowledge about the involved employee.

GRIPE© #6

Be Specific during the Evaluation Process

Overrating (and underrating) employees can cause downstream grief for you, the involved employee, and the organization. There's a tendency to overrate employees because it's easy, and we want our people to do well. Therefore, it's important for you to be specific. If you want to rate an employee as excellent, that's fine so long as you have specific, articulable facts (SAF) to support the high rating.

STRATEGIC HINT FOR YOUR CONSIDERATION

It's acceptable to rate an employee as exceeding expectations or not meeting expectations so long as you have specific, articulable facts (SAF), and these facts are drawn from valid and documented observations.

The same is true with rating an employee as needs improvement. SAFs are drawn from your personal observations, the observations of other supervisors, managers, and executives, and other documentation generated during the rating period. Beware of the common effects that alter sound, clear thinking. These include the halo effect, the harsh and lenient effect, the central tendency effect, the similar to me effect, the first or last impression effect, and the stereotype effect. Each of these effects can cloud the judgment of the evaluator, so it's important you recognize and avoid them.

GRIPE© #7

Remember the Importance of Accuracy

Accuracy is essential in a well-prepared performance evaluation, particularly if the evaluation has negative elements that may impact future employment or promotion. The easiest way for an employee to challenge a negative review is to demonstrate that your specific, articulable facts (SAF) are incorrect. This situation can be prevented by appropriate data collection and analysis during the evaluation process. It can also be prevented by a serious pre-evaluation discussion with the employee where you discuss with him your thoughts regarding his performance and allow him to respond as to the accuracy of your observations. The integrity of the process is dependent on accuracy and objectivity.

GRIPE© #8

Always Proofread Your Documentation

STRATEGIC HINT FOR YOUR CONSIDERATION

If you're being asked to change a performance evaluation to reflect performance traits you know didn't exist, you may be setting up yourself and your organization for bigger problems downstream.

After preparing the document, and prior to taking it to your manager for review, double-check it for accuracy. Minor errors can cause major problems. Your manager will also be looking for errors to prevent such problems. One common mistake on a free-form evaluation process is utilizing a word processor to prepare the document. Please start with a blank screen so there's no possibility that you'll include elements from the last performance evaluation you generated. Also, canned phrases taken from a canned phrase database are easily recognizable and don't reflect well on you as the rater. Anyone can go to a master file of prepared phrases that have passed management scrutiny and cut-and-paste them into this new document. That's the lazy way, and your employees and organization deserve better.

GRIPE© #9

If You're Right, Don't Change It

This rule enters dangerous territory and can put you in an awkward position. You prepared an honest evaluation, and you're directed by your boss (or your boss' boss) to change the ratings either up or down. Before you anchor down, analyze their request and utilize your discretionary time to think it over. They may be right, and you may have made a mistake. Or, they may be utilizing information that was outside your sphere of influence. (Aren't you glad you didn't already have the document signed by the employee?) If the request is in good faith, and is simply a difference of opinion, then it may be in your best interest to follow their guidance. However, if the recommended change is wrong, and against the spirit of your policy, you should make your feelings known. This rule is a personal call and dependent on the strength of your position relative to your boss, but if you change a document to reflect performance traits that you know didn't exist, you may be setting yourself up for bigger problems downstream.

GRIPE© #10

Learn from and Share Your Experiences

STRATEGIC HINT FOR YOUR CONSIDERATION

Performance evaluations straddle the line of providing employees with meaningful improvement and articulation of expectations while also serving as a problem lying in wait if the system isn't properly designed, kept up to date, and fully implemented.

As with any other activity, the more you do it, the better you'll get at it. Share your knowledge with your peers. If your newfound knowledge isn't in the organization's policy, make the appropriate recommendations to get the policy updated to reflect the better way to do evaluations. Finally, performance evaluations may well be the most important document you generate for your employees so take the process seriously and everyone will benefit.

FINAL COMMENTS

The finest people I have met across our great country are in public service like police, fire, and water-related entities. But having good people isn't enough. Give me your best ten people in any job description, and give them a low-frequency activity, particularly one of high risk in nature. Without a policy (system), you'll have ten good people approaching the activity ten different ways with ten different results. Some of these results might be quite bad. Good people need good policy. Performance evaluations require a deliberative and coherent policy. It's a high-risk activity with low frequency, but it's discretionary time. You must use that time to ensure the job gets done right and in conformance with policy.

Performance evaluations straddle the line of providing meaningful improvement and articulation of expectations to employees while also serving as a problem lying in wait if the system isn't properly designed, kept up to date, and fully (and consistently) implemented. The reality is you should do away with performance evaluations, at least the current way you do them. The better solution is to modify your policy and take the activity seriously. A properly prepared performance evaluation is an excellent risk management tool. It's a regular opportunity to assess how a given employee is currently doing and to provide control measures to address risk and elevate future performance. A lot can go wrong with performance evaluations, but you can stay out of trouble if you avoid four things: (1) using the same or similar evaluation year after year; (2) overrating employee performance; (3) rewarding mediocrity; and (4) building a case for retaliation claims. A properly designed performance evaluation system must include: meaningful job descriptions; identified objectives for each job;

a process to ensure employees are meeting these objectives; a process to collect and analyze employee performance data; goals for the next reporting period; and a validated rating system.

That concludes meaningful performance evaluations. Our next stop is the important role of supervisors in achieving organizational mission and purpose through influencing the actions of employees.

Chapter Takeaway on Media Relations/Crisis Communication

Please refer to the media relations/crisis communication material in the Addendum section and ask yourself how you would address the following scenario:

You've just been served with four discrimination lawsuits from protected class employees because their supervisors put them on action plans after their most recent performance evaluation. You're new to the organization and were aghast at the way performance evaluations were handled. Everyone, with few exceptions, received glowing reviews. You knew those findings were wrong because the board hired you to fix this mess and several other troubles of the prior executive. For the last six years, your organization led its peer group in personnel injuries, civil liability, customer complaints, and low productivity. It was evident that mediocrity had subsumed the organization and accelerated its decay. Your first action was to remove mediocrity with accountability, and you did that with tough but fair performance evaluations.

You personally reviewed and signed off on each of the evaluations this year. They all met your tough but fair standard. It upsets you that someone would assume you were discriminating against a protected class employee. All you wanted to do was remove mediocrity and restore accountability. Your standards were universally applied to all employees. Frustratingly, these four plaintiffs are still employees of the organization and poisoning the hard work and control measures you've implemented to improve morale, productivity, injuries, embarrassment, and goodwill. They also are jointly represented by a prominent personal injury attorney who's renowned for leveraging the media against her adversaries. You expect a press conference by the plaintiff counsel tomorrow where she'll outline a litany of aspersions against you and the organization. Exhibit one will be years' worth of glowing performance evaluations. Exhibit two will be copies of the most recent and harsher performance evaluations, including arguments that these evaluations included insufficient details, disputed facts, and punitive action plans not in conformance with prior evaluations. You stand by the decision to change the performance evaluation system, but did you act too quickly? Should you have allowed more time to transition from the old system to the new system? And what about the lack of specific, articulable facts in the evaluations you signed?

It was clear these employees were deficient, but their supervisor hadn't compiled sufficient documentation to cite actual examples. Should you have softened the action plan and scoring without these details? You're also worried the supervisors will taper their comments to deflect from their own performance standards. Will these supervisors vacillate during deposition and assert the new system was too tough, not clear, and forced from above? The entire situation is a mess, but did you make it worse with your haste? You know the plaintiff counsel will weave misinformation, which can only be undone if you confront the propaganda and provide the facts. That will require a press conference, but how do you handle the media while still protecting your rights under litigation? It's a fine line, but you must provide some counter position to protect your goodwill and reputation. What do you do? How will you do it? And what type of remarks (formal or ad hoc) as well as communication style will you deploy?

---

7 Supervisory Performance

Ballast of Mediocrity or Accountability

Photo by David J. McNeil

STRATEGIC HINT FOR YOUR CONSIDERATION

Most tragedies involving water-related entities are avoidable gray rhinos as opposed to black swans or rhinos disguised as black swans. Your tragedies are clearly evident yet there's a tendency to overlook them.

DOI: 10.1201/9781003229087-7

SUMMARY

Supervisors are the catalyst for employee accountability and integrity but also for mediocrity and dishonor. They serve as a bridge between management and line-personnel as well as an influencer of employee behavior and an enforcer of organizational policy (systems).

Supervisors must also articulate the underlying rationale of organizational policies (systems), mission, and purpose in a way that line-personnel can embrace and understand. Most problems lying in wait are rooted in supervisors not behaving like supervisors, or supervisors trying to behave like supervisors but not receiving support from management.

Chapter emphasis will center on ten activity-specific rules for providing meaningful supervisory performance, including ways to improve the selection, quality, mentoring, and development of supervisors.

Hello again. Gordon Graham here, and thanks for continuing our journey toward *real* risk management. The focus in Chapter 7, "Supervisory Performance," continues to be organizational risk management and its five pillars of success. Our emphasis will be pillar four: supervision. We know organizational risk management involves addressing the root causational factors of tragedies (i.e., people, policy, training, supervision, and discipline) and integrating these factors into proactive, resilient systems. By converting problem factors (problems lying in wait) into solution factors, high-reliability organizations attain excellence through the five pillars of success.

Good people who act with good policy, who are regularly trained, properly supervised, and in an organization that addresses arrogance, ignorance, and complacency with prompt, fair, and impartial discipline will prevent tragedies by doing the right thing, the right way every time. That correlates to the fulfillment of organizational mission and purpose as well as maximization of employee fitness and customer loyalty. Each of the five pillars is interrelated and self-reinforcing, but supervision is the pillar with the most influence, or ballast, on an enterprise's behavioral traits and intrinsic psyche.

Let's return to that four letter word, *bias*, for a moment. I occasionally receive evaluations and feedback from my live classes, and the most common complaint from attendees is a bias that I'm a negative person. If I come across as negative, then I failed because I'm positive about water/wastewater operations. The vast majority of things you do, you're doing right. Across this great country, your profession has a lower rate of tragedies than other governmental organizations. My problem is your tragedies are avoidable gray rhinos. They're not black swans or even rhinos disguised as black swans. Your tragedies are clearly evident yet there's a tendency for water-related entities to overlook them. In many cases, your real problems lying in wait are attributed to the root causational factor of supervision. Specifically, a supervisor not behaving like a supervisor, or a supervisor trying to behave like a supervisor but not receiving management support. This problem factor must be addressed and converted to pillar four of the five pillars of success. And that goes back to the confining influence of bias on perspective, mindfulness, and decision making. Bias frequently leads to an incomplete and narrow view of *real* risk management.

Supervision: Selection, Training, Mentoring & Development
➲ People
➲ Policy
➲ Training
➲ Supervision
➲ Discipline

FIGURE 7.1 Supervision: selection, training, mentoring, and development. Source: Graham Research Consultants, LLC.

I already shared the quote from Ross Swope:

> On any given day, nobody knows where the general manager is, and nobody knows where the department manager is, but everybody knows where their supervisor is. And many employees will modify their behavior based on which supervisor is on duty.

Please start studying tragedies in your profession, and you'll see a recurring theme around pillar four. Supervisors influence behavior, and they're responsible for consistent enforcement of organizational systems. Let me accentuate this point. I recently received a phone call at my law office. A chief of police wanted me to teach a class on report writing. "Gordon we need you to do a class on report writing." Why? "I listened to one of your partner's presentations who said three quarters of our liability losses arise from lousy reports." I agree with that. "That's why I'm inviting you to teach a class on report writing." Chief, you're addressing the proximate cause. Let's look for the root cause. The real problems lying in wait. "Gordon, what do you mean?" Well, root cause number one: If you've hired unwise people who can't write reports, then a four-hour report writing class isn't going to turn them into writers. And root cause number two: On all these inadequate reports, is there a supervisor's signature on the report approving its form and content? "Yes." It's a supervision issue. Figure 7.1 summarizes that familiar depiction of the five pillars of success, with emphasis on pillar four and the importance of selecting, training, mentoring, and developing your supervisors.

STRATEGIC HINT FOR YOUR CONSIDERATION

Effective supervision requires proper training, mentoring, developing, selecting, and recruiting of qualified women and men as well as a ten-thousand-hour rule of practice and experience.

Let's accentuate this point of supervisors acting as a ballast of mediocrity or accountability. Recently, there was a story in the Orange County Register about a Department of Motor Vehicles employee, who for years had been sleeping on duty for several hours a day. Her supervisors were aware of the behavior, but nobody wanted to address the issue. That's a failure of supervision, and the result was organizational degradation by mediocrity. How do you recruit, select, train, mentor, and develop the next generation of supervisory women and men who will do their job and enforce your policies (systems)? Let's break these questions down. How do you select your supervisory team? I know there are different size organizations reading this book. I also imagine there are different processes that vary from state to state. Generally, I see the following in water-related entities: You announce a test, you give a test. You establish a list. You select people off the list. For some of you, you're saying: "Gordon, that's exactly how we do it."

See, you take this woman off the list, you make her a supervisor. A month into the promotion, she figures it out.

> I have to challenge, confront, and honestly evaluate people? I have to tell people not to do things I used to do with them when we worked together? I don't feel comfortable telling people they made a mistake. I don't think I can honestly evaluate my people because they're friends. I had no idea what this supervisory job was all about.

So what does she do at that point? Does she come up to her manager and say, "I made a huge mistake in this promotional process. Please give me my old job back and make me look inept." No. She'll stay in grade for how long? Forever. And the best we receive is mediocrity. Once mediocrity is present, accountability leaves the organization. Without accountability, the organization begins to degrade. Line-personnel reporting to this supervisor will start to do what they want rather than follow policy. That deviation from policy will have a cascading effect throughout the organization. Sooner or later, something bad will happen. Show me a tragedy in water/wastewater operations, and I'll show you a proximate cause of X. The real problem lying in wait, however, is all too often a supervisor not behaving like a supervisor. Or, alternatively, a supervisor who tried to behave like a supervisor but wasn't supported by management.

Could that describe your organization? Do you have the necessary training and mentoring processes to prepare personnel to take on leadership roles and support them after promotion? Maybe there's a better way to promote people. "Gordon, if you were the executive-in-charge, how'd you do it?" Thought number one: the ten-thousand-hour rule. Malcolm Gladwell is all over my recommended reading list. He has several great books, including *Outliers*. I read his ten-thousand-hour rule in *Outliers*. In order to master any profession, you have to do it for ten thousand hours. He provides details of athletes, artists, musicians, and others who reached the pinnacle of their profession through innumerable hours of practice. Apply the ten-thousand-hour rule.

STRATEGIC HINT FOR YOUR CONSIDERATION

Executives have a key role in promoting the next generation of women and men to enforce the organizational rules you've put in place. That's the legacy of executives.

When I was active at the California Highway Patrol, you could promote to sergeant first-line supervisor with two years on-the-job experience. You haven't even learned how to be a cop yet, and we're going make you a supervisor. That's a problem lying in wait. Ten-thousand-hour rule. I recommend a minimum of five years organizational experience as well as certification requirements before people are eligible to take promotional tests. Filter out the people who are promoting for the money through experience and certification requirements. Many professions require supervisors to have certificates on human resources (HR) management and conflict resolution before they're eligible to promote. I'm aware your established process may be mandated by civil service rules designed to be objective and fair, but this approach is out of date and a ticket to tragedy. There are better ways to promote people. You may also encounter resistance from your HR personnel who'll impart all the reasons you can't put a more efficient and effective process into place. Notwithstanding these issues, advocate the ten-thousand-hour rule as well as certification requirements.

What have you told your people about identifying those women and men who will be future supervisors in your organization? Or, are you simply hoping the best-of-the-best will apply for these supervisory positions? After you recruit and encourage great people to promote, what are you doing to prepare them for the testing process? Do you have study groups to help them understand the process, the testing components, and the preparation requirements? You may incur some internal clashes trying to implement these standards, but you should seriously consider the recommendation. Make sure your promotional tests are related to the job they're going to be doing as a supervisor.

Executives have a key role in promoting the next generation of women and men to enforce the rules you've put in place. This next generation will develop the future policies and procedures of your organization long after you've retired from the organization. That's your legacy. With respect to testing, does your test measure what's required to be a good supervisor, or does it test for rote memory? Is an ability to recite all the details of a low-risk, high-frequency activity really the standard by which you want your future supervisors measured? After they've been promoted, do you have a training program (external or internal) for new supervisors while they're on probation? That's an important step to ensure they've the necessary knowledge, skills, and abilities to do their new job. Finally, have you considered bringing back the best-of-the-best to help further develop your new supervisors? Those who are executives didn't just end up as a senior leader. They worked hard and learned from those who stepped up and guided them. I had the greatest sergeant in the history of my organization, the California Highway Patrol. He knew everything about how to be a great sergeant, but when he retired, he took his one-million memory markers with him. Why don't we identify these great women and men retirees and bring them back to impart their institutional knowledge to the newest generation of supervisors? Remember, precept number one: If we don't understand our past errors, we'll continue to repeat these errors in the future.

STRATEGIC HINT FOR YOUR CONSIDERATION

Poor supervisors still move organizations in the right direction, albeit not quickly enough. Failure to supervise moves organizations backward because there's no sitting still in your high-risk and complex profession.

Sometimes, supervisory failure is situational. A supervisor may be having a bad day or might be sick or fatigued. On a deeper level, being a supervisor can be uncomfortable. It requires you to ask whether you're really doing everything you're supposed to be doing. And sometimes the answer is difficult to face. As a result, you're not sure what decision to make so you don't make one at all. On an individual level, fighting the failure to supervise means taking the time to care. It's important to distinguish between failure to supervise and ineffective supervision. Poor supervisors still move organizations in the right direction, albeit not quickly enough or with needed intensity. Failure to supervise regresses organizations because in high-risk professions like water/wastewater, there's no sitting still. Bottom line: Failure to act retrogrades organizations and their personnel.

There are three main factors contributing to supervisory failure within water-related entities: (1) Lack of mentoring before the supervisor is promoted: No one showed her how to lead. It's a lot easier to replicate the needed behavior when you've seen it in action. (2) Wrong people in supervisory positions: Water/wastewater professionals who are book smart and can promote quickly may lack practical skills to take decisive action. Since most water-related entities are adequately operated and sourced with professional line-personnel, it takes longer to expose this deficiency. And by the time it's obvious, the ability to reassign may have passed. (3) Burnout: Sometimes, failure to supervise is seen in excellent supervisors who are capable people with a track record of turning around problem personnel and fixing problems. This reputation sometimes leads to more assignments and more problems until they become burned out and give up. At one time they led, but now they're just getting by.

STRATEGIC HINT FOR YOUR CONSIDERATION

Mentoring is a core component of development and warrants bringing back retired leaders who can offer directional guidance for new supervisors and the boosting of memory markers and perspective.

STRATEGIC HINT FOR YOUR CONSIDERATION

Lack of supervisory leadership means line-personnel don't understand the why of the work they do, and, as such, will lose their enthusiasm as well as their resolve to do the right thing, the right way every time.

Even more dangerous is when a failure of leadership permeates the organization and becomes the norm rather than the exception. Then, additional consequences (tragedies) arise, including increases in water/wastewater personnel injuries and on-the-job deaths. Strong organizational leadership is needed because water/wastewater operations are inherently risky. When supervisors are reluctant to make decisions, the risks are exacerbated. Lack of supervisory leadership means line-personnel don't understand the why of the work they do. Supervisors must articulate the mission in a way that's easy to understand and get behind. When they fail to do so, line-personnel will lose enthusiasm for the work they do and the perseverance to do the right thing, the right way every time. A rogue line-personnel is often the result of a weak supervision. These individuals frequently have noble intentions and are driven to do the right thing, but no one is showing them the path so they forge their own way. Even in organizations with strong supervision, there will always be employees who operate outside organizational culture. These exceptions are always fewer in number and less impactful in high-reliability organizations.

Supervisors can't grow and develop in a culture devoid of leadership. As new supervisors move up in the organization, they're likely to emulate behavior. Oftentimes, the only way to break this cycle is to bring in leadership from outside the organization. So how do we fight leadership failure? It involves empowering employees and encouraging them to have a stake in the overall mission. Perhaps counterintuitively, it means setting high expectations for line-personnel and supervisors. At the same time, your organization must support personnel as they promote up. You must provide professional development resources at all levels, including educational opportunities and operating skills. These programs include invitations to participate in the next level up before being promoted so employee suitability for the position is assessed. On an individual level, fighting the failure to supervise means taking the time to care. Your people are your best resources, but too often your actions don't support that. Equally important, know your organization's policies. What does your organization expect? When you see people operating outside those expectations, get them on track using passive or active discipline. Supervising requires courage. It's not hard, but it requires knowing what to do. The best balance is a combination of measured discipline and a desire to do what's right. Every day you don't act is one day closer to tragedy.

Mentoring is important in any organization. Many organizations have formal or informal mentoring programs to help new people succeed. But what about after that? Is that when mentoring stops? Hopefully not. Mentoring is a way to supplement basic training. Learning in the context of a relationship is what mentoring is all about. Mentoring can help people learn valuable skills more deeply, and is effective for probationary employees as well as new supervisors and managers. Sharing experience is important because we've all made mistakes, and each is an opportunity to learn. By sharing experiences, you can help others avoid similar mistakes. Whether it occurs in a formal or informal setting, mentoring can be one of the most valuable professional relationships we experience. Supervisors act as the epoxy for organizational risk management, and the foregoing examination underscores the necessity to enhance supervisory performance through a comprehensive and deliberative process of rules and systems. I created ten activity-specific techniques to improve supervisory performance, which I call GRISP© (Graham's rules for improving supervisory performance). These rules encompass core areas of meaningful supervision that executives, managers, and supervisors should know. My system improves supervisory quality and prevents consequences (tragedies) arising from improper behavioral traits. Figure 7.2 summarizes these rules. Let's examine each in detail.

Graham's Rules for Improving Supervisory Performance (GRISP©)

GRISP© #1	➲	How do you select your supervisory team? This process is a management and human resources issue, but here are some thoughts for you.
GRISP© #2	➲	Some initial training considerations.
GRISP© #3	➲	Some ongoing training considerations.
GRISP© #4	➲	Clearly define your expectations of performance.
GRISP© #5	➲	Share with your supervisory team the problems caused by a lack of system implementation and system enforcement.
GRISP© #6	➲	Hey Boss! Back up your supervisors when they make the tough calls.
GRISP© #7	➲	You must have a feedback loop to ensure what you say you're doing is in fact being done.
GRISP© #8	➲	Give your people credit for what they've done.
GRISP© #9	➲	Constantly be looking for the next best way.
GRISP© #10	➲	A thought from Ross Swope.

FIGURE 7.2 Graham's rules for improving supervisory performance (GRISP©). Source: Graham Research Consultants, LLC.

GRISP© #1

How Do You Select Your Supervisory Team?

Apply the ten-thousand-hour rule on the job requirement. There should be certification requirements prior to promotion as well as advanced testing considerations and preparations. Supervisory applicants should receive interim responsibilities to assess suitability. You must take the new supervisor probationary process seriously. If a newly promoted supervisor isn't doing an adequate job while on probation, address this issue immediately.

STRATEGIC HINT FOR YOUR CONSIDERATION

Supervise by walking around (SBWA) is imperative. Catch your people doing something right and document it. Line-personnel frequently only hear from their supervisor when they do something wrong.

GRISP© #2

Some Initial Training Considerations

Training must commence prior to supervisory operations and should encompass buddy-to-boss considerations as well as daylight-versus-artificial light decision making scenarios. New supervisors must be taught how to think through their activities and the importance to maximize discretionary time when available. They similarly must know organizational policy to enforce it. The utilization of mentors by enlisting best-of-the-best will provide memory markers that will accelerate the development of your supervisors.

GRISP© #3

Some Ongoing Training Considerations

Continuous learning via mandatory reading programs is essential. There's also value in learning from good and bad events. We must learn from mistakes prior to tragedy, which necessitates a program for non-punitive, close-call reporting (NPCCR). My thesis in grad school was on the mathematical relationship between close calls, mishaps and tragedies. The work of H.W. Heinrich in the early 1900's is valuable. Learning from deaths and major injuries is a good idea. The better idea would be to learn from the mishaps - the sprains, cuts, bruises, falls, property damage only events. The best idea is to learn from the "close calls." They are much more frequent and likely are not being captured in your reports. Firefighters are actively learning from these "close calls" and have implemented the following websites where such knowledge can be shared: www.firefighterclosecalls.com and firefighternearmiss.com. A similar initiative should be implemented within the water/wastewater profession.

GRISP© #4

Clearly Define Your Expectations of Performance

The primary mission of a supervisor is system enforcement. Someone has to enforce your organizational policies and procedures. Make every day a training day and focus it on core critical tasks (non-discretionary time) and critical tasks (discretionary time). Genuine concern for your personnel is paramount. Supervise by walking around (SBWA) is imperative, and catch your people doing something right and take the time to document it. The most frequent complaint I hear from line-personnel in every profession is they only hear from their supervisor when they did something wrong. Please think this one through and document observed good behavior regularly as well as communicate the rationale behind organizational systems and the importance of employee conformance. Your employees deserve to be acknowledged for following the rules. The stick of supervisory enforcement must be balanced with the carrot of praise.

STRATEGIC HINT FOR YOUR CONSIDERATION

The adage of perception being reality has validity. If supervisors think they won't get the full support of their manager after a tough call is made, they'll avoid making the tough calls.

GRISP© #5

Share with Your Managerial and Supervisory Teams the Problems Caused by a Lack of System Implementation and System Enforcement

When rules aren't followed, bad things can happen. These include death, injury, embarrassment, civil liability, internal investigations, and even the rare criminal filing against your personnel. Knowledge and support behind the underlying rationale of organizational systems are needed to foster employee support and enforcement for such systems.

GRISP© #6

Hey Boss! Back up Your Supervisors When They Make the Tough Calls

The adage of perception being reality has validity. If supervisors think they won't get the full support of their manager after a tough call is made, they'll avoid making the tough calls.

GRISP© #7

You Must Have a Feedback Loop to Ensure What You Say You're Doing Is in Fact Being Done

Formal audits are essential but so are informal audits like managing and supervising by walking around (MBWA and SBWA). It's also important to analyze activities that ended up right because activities ending up right doesn't mean you did things right.

GRISP© #8

Give Your People the Credit for What They Have Done

Prepare them for further promotion, reinforce the value of their contribution, praise them in public, and criticize in private. Set the proper example.

GRISP© #9

Constantly Be Looking for the Next Best Way

There's always a better way, and we must be searching for it. The importance of sharing memory markers can't be overemphasized.

STRATEGIC HINT FOR YOUR CONSIDERATION

"The major cause in the lack of integrity in American Police Officers is mediocrity."

That statement from Ross Swope applies to any organization, including water-related entities.

GRISP© #10

A Thought from Ross Swope.

In the 1990s, Los Angeles Police Department (LAPD) was chastened by the Rampart Scandal. After the fact, a number of experts (mostly lawyers) analyzed what caused the tragedy. While the lawyers focused on proximate cause, one commander identified the real problem lying in wait. His name was Ross Swope, and the essence of his comments is good for you to know: "The major cause in the lack of integrity in American Police Officers is mediocrity." That statement applies to any organization, including water-related entities.

FINAL COMMENTS

The fourth pillar in the five pillars of success is the key role of the supervisor. Show me a tragedy in water/wastewater operations, and I'll show you a proximate cause of X. The real problem lying in wait, however, is all too often a supervisor not behaving like a supervisor. Or, alternatively, a supervisor who tried to behave like a supervisor but wasn't supported by management. Either of these scenarios is a problem lying in wait and requires your attention. The key role of supervisors is systems enforcement, and their actions are pivotal to the success of new policies or procedures (systems). As a role model and mentor, supervisors lead by example. Those who wince, raise an eyebrow, and blatantly criticize new policies will lead to its disregard by line-personnel. Conversely, a supervisor who conveys the positive aspects and benefits of new policies will ensure its successful implementation.

When it comes to safety, the supervisor must be the enforcer to guarantee compliance with safety regulations. A capable supervisor produces meaningful performance evaluations with the goal of improving employee safety and providing opportunities for promotion. They often know when a mistake is made from a training failure rather than an act of mal-intent. Supervisors know when to give a pep talk, when to add humor to lighten a mood, and when to be deadly serious when confronting a dangerous situation. An accomplished supervisor understands the organizational dynamics and commands respect from her direct reports.

STRATEGIC HINT FOR YOUR CONSIDERATION

A competent and compassionate supervisor can make a tough job fun for line-personnel, but a deficient supervisor can have a devastating effect on employee morale and organizational psyche.

The supervisor is similarly the linchpin between the community, management, and line-personnel. They have a direct effect on morale (positive and negative), including overall organizational psyche. Supervisors are responsible for the safety, discipline, and training of each employee under their watch. They're the mediator between line-personnel and management and are charged with ensuring that organizational personnel are safe, represented, and proficient. Without question, the supervisor is the most important leader in a water-related entity. A competent and compassionate supervisor can make a tough job fun for line-personnel, but a poor supervisor can have a devastating effect on morale and operations. Supervisors are the organization ballast of mediocrity or accountability.

We have three chapters left, and their topics will comprise employee discipline, job-based harassment, and my favorite topic: ethical and sound decision making. Please remember the five pillars of success: I believe you have good people. I know you can get good policy. Let's make every day a training day and focus it on core critical tasks and critical tasks. Let's get supervisors behaving like supervisors where there's consistent and universal enforcement of organizational policy. On those rare occasions when people don't follow rules, there must be measured discipline to address the issue, and it always must be a function of policy and not consequence. Until then, please work safely.

Chapter Takeaway on Media Relations/Crisis Communication

Please refer to the media relations/crisis communication material in the Addendum section and ask yourself how you would address the following scenario:

Your field superintendent just left your office. A line-personnel was involved in a major at fault automobile accident involving a pregnant mother of two. You asked for details and are overwhelmed with sadness and disappointment when you hear them. There was a minor valve malfunction that was spewing water into the air. One of your supervisors asked Phil, a distribution operator, to fix it as soon as possible. You knew from experience that Phil had a history of speeding to such malfunctions and breaks. One time, he had a close miss while driving with an employee to a minor water line break. Another time, he was issued a citation for speeding through a yellow light. This time, he ran a red light and collided with the pregnant mother's vehicle in an intersection. The police report details the accident, and it makes you shiver.

Before leaving your office, the superintendent advised Phil's MVR was pulled six months ago, and there were three moving violations over a two-year period. You ask what measures the supervisor had taken, and the answer was unacceptable. There were several verbal warnings but no written documentation and no action plan. When asked if the supervisor had other line-personnel that weren't following policy or meeting employment requirements, the superintendent confirmed there were two other direct reports with similar driving records and policy violations. This supervisor was a high-performing distribution operator before promoting two years ago. You sense an inability by the supervisor to act like a supervisor, and you wonder if your superintendent is also part of the problem by not backing this supervisor in tough situations and providing important mentorship.

Phone calls are inundating the switchboard from local news media. What happens if the root causational factors behind this tragedy are unearthed. There's no excuse or defense for such organizational ineptness. Your organization could've prevented this tragedy if there had been proper supervisory performance. What do you do? How can you address these issues both internally and to the public? Is it better to own the tragedy and commit to doing better with supervisory selection, mentoring, training, and support? Or, should you not engage the media and redirect them to your general counsel? How can you apply a measured approach that acknowledges the employee error while also detailing organizational steps you're taking to mitigate future similar tragedies?

---

8 Employee Discipline

The Ignored and Inoperative Pillar

2019 Photo Courtesy of Las Virgenes Municipal Water District

SUMMARY

Discipline is the most ignored and inoperative of the five pillars of success. Its frequent disregard impairs the design of viable control measures (systems) by decommissioning the other four pillars and reverting them back to root causational factors or problems lying in wait.

Discipline is a mechanism to change employee behavior and must be applied as a function of policy and not consequence. Organizations should utilize human resources personnel in this discretionary time activity as well as apply documentation, specific articulable facts, and professionalism during the disciplinary process.

Chapter emphasis will center on a series of ten rules intended to impart directional guidance in providing prompt, fair, and consistent discipline as well as their use in bolstering employee knowledge and conformance of organizational systems.

DOI: 10.1201/9781003229087-8

Hello again. Gordon Graham here, and welcome back to our journey toward *real* risk management. We'll continue our focus on organizational risk management and its five pillars of success. Chapter 8, "Employee Discipline," centers on the fifth pillar. We know good people who act with good policy, who are regularly trained, properly supervised, and in an organization that addresses arrogance, ignorance, and complacency with fair and impartial discipline, will prevent tragedies by doing the right thing, the right way every time. The functionality of the five pillars of success is predicated on its interrelated and self-reinforcing dynamics. Simply stated, you can't ascend to a high-reliability organization without fulfilling all five pillars of success.

Discipline is the pillar most often ignored and frequently rendered inoperative. Its frequent disablement divests the other pillars of their utility and purpose, which subsequently impairs an organization's capacity to build actionable control measures (systems). Such systems are leveled ineffective because the root causational factor of discipline isn't properly converted to a pillar of success. The resulting outcome is an uncoupling of the other pillars and their reconstitution from solution factors to problems lying in wait or precursors to tragedies.

For purposes of this chapter, let's assume you have good people working for you. To be successful in water/wastewater operations, good people need good policy, but having good people and good policy aren't enough. Your personnel must also know and understand your policies. That brings up the training component, and the importance of systems training through daily training bulletins on core critical tasks and critical tasks. The correlation of training is discipline, and this pillar can't be taught but can be emulated. It must be an intrinsic part of your organizational culture. Discipline isn't a function of how an activity concluded, but a function of whether policy or procedure (system) was followed. While people, policy, and training are all important, if supervisors aren't enforcing the rules, that's a problem lying in wait. Discipline has a direct impact on the other pillars and must *not* be decommissioned. Figure 8.1 summarizes the five pillars of success with an emphasis on employee discipline.

STRATEGIC HINT FOR YOUR CONSIDERATION

Rules without enforcement lead to impotent policies. There must be consequences for personnel who suffer from arrogance, ignorance, or complacency as well as those who are convinced the rules don't apply to them.

Pillar Five: The Importance of Employee Discipline
Failure of Policy and Not Consequence

- People
- Policy
- Training
- Supervision
- Discipline

FIGURE 8.1 Pillar five: the importance of employee discipline. Source: Graham Research Consultants, LLC.

My focus in the last chapter was the key role, or catalyst, of the supervisor to warrant your employees do the right thing, the right way every time. Supervisors direct and influence line-personnel behavior. The difference between mediocrity and accountability, and achieving strategic objectives, is the effectiveness of your supervisors. High-reliability organizations employ supervisors who enforce the rules because it's the right thing to do, and it embodies their culture. Discipline is the adhesion to ensure employees understand, embrace, and support organizational systems as well as the underlying rationale behind these systems. Rules without enforcement are meaningless and lead to impotent policies (systems). There must be consequences for personnel who suffer from arrogance, ignorance, or complacency, as well as those who are convinced the rules don't apply to them. I'm confident you understand the importance of discipline. The problem I often see in water/wastewater operations is discipline has become a function of outcome or consequence (tragedy) and not a violation or a deviation of policy. Let's examine that point further.

You already know my core thinking on this pillar. Discipline isn't a function of outcome (how the activity concluded), but rather a function of policy (whether systems were followed). Just because an activity ends up well doesn't mean the activity was performed correctly. Maybe you were lucky, and you can't rely on luck. I'll take luck when I get it, but you can't rely on luck. You must rely on systems (i.e., control measures, policies, processes, procedures, rules, checklists, and protocols), and you can't overemphasize the favorable impact of discipline on high-reliability organizations. If your efforts at disciplining an errant employee are unsuccessful, it can cause problems downstream. So why do our efforts at discipline fail? More importantly, what can be done to make discipline prompt, fair, consistent, and impartial? And, how can you guide personnel to take organizational policies (systems) seriously? Now's a good time to reflect on Tony Kern's great book *Darker Shades of Blue*, which tells the story of Czar 52. My purpose for having you read Tony's book is simple: If it's not documented, it didn't happen. The involved pilot had a history of reckless behavior that everyone knew about, but the behavior was never documented and ultimately resulted in a major tragedy. The pillar of discipline was disabled.

STRATEGIC HINT FOR YOUR CONSIDERATION

Discipline isn't a function of outcome but a function of policy because an activity that ends well doesn't mean the job was performed correctly or conforms to organizational control measures (systems).

Let's advance the discussion by reviewing a discipline case. You have an employee who did something wrong. You initiate an effort to discipline this employee, who we'll name Pat (short for Patricia or Patrick). Pat retains an experienced employment attorney to defend him/her at the proceeding. The lawyer will listen to Pat's side of the story and review your allegations as to why he/she needs to be disciplined.

One of the first things the lawyer will do is verify your actions are consistent with any memorandum of understanding (MOU) your organization has with the bargaining group as well as any laws in your state that protect governmental personnel (if applicable). From personal experience, many discipline cases fall apart because the investigator acted in a manner inconsistent with the MOU or wasn't compliant with organizational policy. Having studied many tragedies in water/wastewater operations, I frequently see the same thing: Everyone knew and no one did anything about it. Let's assume your organization followed the applicable procedures, and the discipline case is moving forward. If I were the defense lawyer, my first attack on your case would be the following: Please provide me a copy of the policy that Pat violated. If you don't have a policy, that may be the end of your disciplinary efforts. If your response is, "We don't have anything written down, but everyone knows you're supposed to do X," your case is finished.

Let's further assume you have a properly designed, up-to-date policy, and you present it to me. Here's my second attack: Show me that Pat has been trained on this policy. If your response is, "Here's a document from point of hire that Pat signed acknowledging receipt and review of our policy manual," that's not good enough. You'll have to document a training program, both initial and ongoing, to prove Pat was familiar with the policy. Continuing on, let's assume you have a well-written policy and proof Pat was aware of this policy through initial and ongoing training. My third line of attack will be: Are your efforts to discipline Pat consistent with past efforts when other employees violated the same policy? If I can demonstrate other employees (either in the past or involved in this incident) were treated differently (meaning lesser punishment or no punishment), that will not bode well for your case. If you've done all these things right, meaning you have a well-written policy, you have provided Pat with initial training, you produce proof of ongoing training, and your disciplinary efforts are consistent across incidents and employees, there's still one final step you must provide: documentation.

STRATEGIC HINT FOR YOUR CONSIDERATION

Supervisors are responsible for making sure everyone does the right thing, the right way every time, and they must address violations of organizational policy with discipline. The adhesion for accountability is discipline.

Once again, let's go back to Czar 52. The involved pilot had a long history of inappropriate behavior without negative outcomes. The behavior was known to his superiors and some talked to him about it, but no one documented the misconduct. When new superiors came to the airbase and read the personnel jackets of the pilots, there were no documentations. You know the rest of the story. Having studied many tragedies in water/wastewater operations, I frequently see the same thing: Everyone knew and no one did anything about it. Remember the five pillars of success: people, policy, training, supervision, and discipline. I'm hopeful you have good people and good policy in your organization. Please take the training of your personnel seriously. Make sure you're issuing daily training bulletins and documenting their completion by line-personnel. Your supervisors must take their jobs seriously. Although it's never enjoyable, supervisors must address violations of organizational policy through discipline. You're in a complex, high-risk profession, and you're responsible for making sure everyone does the right thing, the right way every time is important. Discipline is the adhesion for ensuring individual and organizational accountability.

Whenever we discuss discipline, a fundamental question must be broached and remain at the fore: What's the purpose of discipline? In theory, you know the answer: to change behavior. The importance of this question and its answer can't be overstated. When you apply theory to practice, real names are attached to real-life situations. It's common to shift away from the purpose of discipline, which is changing behavior, to doling out an appropriate penalty. In the context of theory, you know it's not about the penalty but changing behavior. In practice, it's about proving misconduct and assessing a penalty. Keep asking yourself that question whenever you have to make a discipline-related decision. Your focus must be on changing behavior and not dispensing punishment.

STRATEGIC HINT FOR YOUR CONSIDERATION

Discipline involves a twofold decision process: (1) whether an offense has been committed and (2) what penalty should be dispensed. Your focus must be on changing behavior and not doling out punishment.

When a supervisor plans to reprimand a subordinate, she's concurrently serving as investigator, witness, judge, and jury. Discipline involves a twofold decision process: (1) whether an offense has been committed and (2) what penalty should be dispensed. As a new supervisor, do you have any training on how to conduct an investigation, make a determination, and decide an appropriate penalty? Your focus must be on changing behavior and not doling out punishment. There's also the issue of consistency in your disciplinary decision making. Is it consistent with your past decision making, and do you believe you can remain consistent going forward with other line-personnel on similar facts? Equally important, is your decision making consistent with other supervisors in your organization who've handled similar situations? The need for consistency transcends the appropriateness of a particular penalty.

Many water-related entities are classified as governmental employers so you must be attentive to principles of due process and just cause. Both require an employer to be consistent throughout the disciplinary process. Disciplinary decisions, such as when matters are investigated, who conducts the investigation, how the investigation is carried out, the burden of proof applied, and the penalty imposed, must be performed in a consistent manner. The importance of consistency necessitates your strong consideration in bifurcating the administration of discipline from the reporting of misconduct. This division involves removing the supervisor from the former since she's responsible for the latter. Once misconduct is reported, your organization's human resources personnel and executives should be responsible for ensuring consistency in the investigative process and disciplinary outcome. It's also important to remember that while some offenses are unacceptable by their nature, most are not, and progressive discipline may be the prudent choice. Ask yourself the following questions: If my manager received an identical complaint about me, would a reprimand without prior warning be necessary to change my behavior? Might my behavior change if my manager made me aware of it? Remember the golden rule: Treat others the way you would want to be treated.

High-reliability organizations are known for doing the right thing, the right way every time because they're aware of and pay attention to the five pillars of success. These pillars are getting and keeping good quality people, developing and maintaining good organizational policies, assuring personnel are adequately trained, having supervisors who'll enforce organizational rules, and applying discipline when rules aren't being followed. Take away one of these pillars, and bad things will happen. They're inextricably interlinked in the support of your mission, purpose, and success. The fifth pillar, discipline, is often disabled, and that leads to tragedies. What can be done to prevent problems in this regard? You can utilize a system of rules that provide directional guidance on the importance of employee discipline. Allow me to introduce GRAED©, which is my system for appropriate employee discipline. Figure 8.2 summarizes these rules. Let's examine each in detail.

GRAED© #1

Understand the Importance of Organizational Discipline

If you want things to go right, you must be aware of the five pillars of success. These pillars are getting and keeping good quality people, developing and maintaining

Graham's Rules for Appropriate Employee Discipline (GRAED©)

GRAED© #1 ➲ Understand the importance of organizational discipline.
GRAED© #2 ➲ Develop good organizational policies and procedures (systems).
GRAED© #3 ➲ Make sure people know the policies (systems) and your performance expectations.
GRAED© #4 ➲ Make sure your supervisors are behaving like supervisors.
GRAED© #5 ➲ When rules (systems) aren't followed, someone needs to act.
GRAED© #6 ➲ Take the time to think it through, and keep it private.
GRAED© #7 ➲ Be specific when talking to the involved employee.
GRAED© #8 ➲ Keep it professional, fair, and balanced.
GRAED© #9 ➲ Documentation is essential.
GRAED© #10 ➲ Get on with life and your organization's mission.

FIGURE 8.2 Graham's rules for appropriate employee discipline (GRAED©). Source: Graham Research Consultants, LLC.

good organizational policies, assuring personnel are adequately trained, having supervisors who'll enforce organizational policy, and applying discipline when rules are violated. Take away one of pillar, and bad things will happen. They're inextricably intertwined in supporting your organization.

GRAED© #2

Develop Good Organizational Policies and Procedures (Systems)

STRATEGIC HINT FOR YOUR CONSIDERATION

Take away one of the five pillars, and bad things will happen. They're inextricably interlinked in the support of your mission, purpose, and success. The fifth pillar, discipline, is often disabled, and that leads to tragedies.

Why do we need employee discipline? It's part of the five pillars of success. If you're going to be successful in the discipline process, it starts with building good policy manuals (systems). You must hire good people and give them good policies. As I travel around this great country, and as I talk to people and view their policies, I regularly see missing policies, no policies, out-of-date policies, illegal policies, unconstitutional policies, inept policies, and inconsistent policies. You have to make sure your policies are effective.

Let me accentuate this point with an example. I'm marketing my company, Lexipol, which is an entity I started twenty years ago to standardize policies, procedures, and training for public safety organizations. I'm proud that thirty-eight of our great states use the Lexipol knowledge management system. In my pitch, a Delaware cop puts his hand up. "Gordon, we're thinking about using your policy manual in our police department, but I have a question: How many pages of policy do you have?" And I gave him the answer. We've one-hundred-sixty policies in the Lexipol policy manual. His retort: "That wasn't my question. My question was how many pages of policy do you have?" Well, no one has ever asked me that question so I sent a text message back to Lexipol headquarters and received a prompt response: "Mr. Graham, we've four hundred pages of policy in the manual." So I told the cop that we have four hundred pages of policy in the manual. Do you know what he says to me? "That's not enough." What do you mean it's not enough? "We currently have twelve hundred pages of policy in our manual. Four hundred isn't enough." He was serious. "We have twelve hundred pages of policy. Four hundred isn't enough." And I said, do you really have twelve hundred pages of policy? "Yes sir."

I know a police department that has twelve thousand pages of policy manual. They also have a document signed by every employee certifying they've read and understand these policies and will be accountable for applying them. That's a lawyer trick. Nobody knows twelve thousand pages of policy. I don't care how many pages

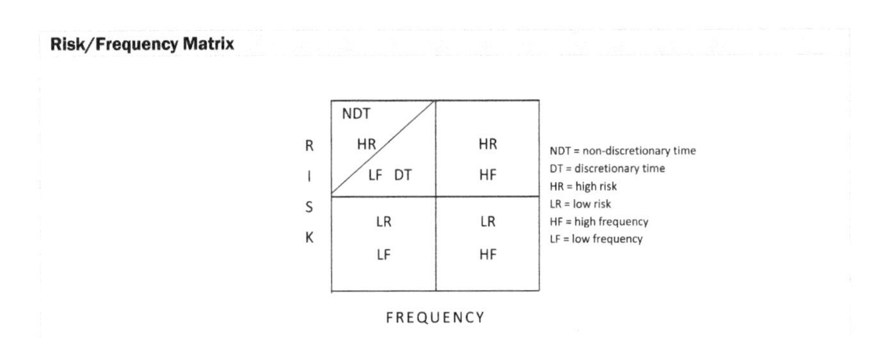

FIGURE 8.3 Risk/frequency matrix. Source: Institute of Safety & Systems Management.

of policy you have in your manual. That's not the issue, but I want you to start looking at your policy manual. Let's go back to our familiar risk/frequency matrix, which is one of the most important investigative tools you have along with risk assessments, to properly recognize and prioritize the thousands of risks facing your organization. Risk/frequency matrixes are your training precursors to achieving pre-incident knowledge level verification on activities that are high risk, low frequency, with no time to think (core critical tasks) as well as activities that are high risk, low frequency with time to think (critical tasks). Figure 8.3 is a reproduction of this important tool.

STRATEGIC HINT FOR YOUR CONSIDERATION

Color code your policy manual so you can easily identify what's important for any job description. White pages are nice to know. Yellow pages require familiarity. Red pages are absolutely needed to know.

The police department I referenced above has twelve thousand pages of manual, but eleven-thousand-five-hundred pages comprise the bottom two boxes on the risk/frequency matrix. They're insignificant, low-risk administrative activities. There's five hundred pages encompassing the top two boxes, and it's those activities that lead to tragedies. Please start looking at your policy manual using the risk/frequency matrix and color code your policy manual. "Gordon, what do you mean by that?" Print all your low-risk policies that are in the bottom two boxes on white paper. "Gordon we've done that." Good for you. Print your high-risk, high-frequency, with no time to think and your high-risk, low-frequency with time to think (critical tasks) policies on yellow paper. And print your core critical task policies, which are high risk, low frequency, with no time to think (the top-left box, upper left corner) on red paper. Color code your policy manual so you can easily identify what's important for any job description. White pages are nice to know. Yellow pages require familiarity because you're doing it a lot or you've time to think. Red pages are core critical tasks: absolutely need to know, and, again, the necessity of daily training. More on that later.

Another thought on pillar two. Policies are like painting the Golden Gate Bridge, which is a work in progress. My hometown is San Francisco, and you know what I've learned? Every day, they're painting the Golden Gate Bridge. When they finish painting it, they do what? They start painting it again. Whenever I get a phone call from an executive in any organization that says: "Gordon, we took your advice. We redid our policy manual, and it's finished." Then you didn't take my advice. It's never finished but a continual work in progress. All you are is a temporary custodian of that policy who's keeping it up to date during your tenure. Time for a refresher. You're disciplining an employee for misconduct. You might even be trying to fire the employee for misconduct. He comes to my law office, and I sign him up as a client. Your employee is now my client. I'm representing my client in front of you. Here's my first attack: What did my client do that's the genesis of this discipline case? "Well he did this, this, and this." My second attack is show me the policy. Show me the written policy that prohibits the behavior. And if your response is, "Well, we really don't have a written policy but everybody knows it," then your case is over. You won't be disciplining my client, your employee, without a written policy.

STRATEGIC HINT FOR YOUR CONSIDERATION

New policies and procedures must be taught to employees regardless of how long they've been in a job, and you may find it helpful to introduce such policies at the same time (annually) to ensure proper training of your personnel.

GRAED© #3

Make Sure That People Know the Policies (Systems) and Your Performance Expectations

Having good people and good policies aren't enough. The third pillar of success is training. Initial training is essential. Make sure your people know the policies and your expectations for their performance. You can have excellent written policies and procedures, but they're ineffective without documentation proving your employees were adequately trained on them. Lack of documentation is a problem lying in wait. Equally concerning, you can't effectively discipline without verification of ongoing training on policies that were violated. I'm pretty sure you're taking the initial training process seriously. My concern is ongoing training. After people are hired and released from probation, when's the next time they have to take a serious test with advanced preparation? The answer for so many organizations is they don't. Initial and ongoing training must emphasize core critical tasks. Every day is a training day, and your focus must be on those activities that have the highest probability of ending up in tragedy. Identify the aforementioned activities that are high risk, low frequency, with no time to think (core critical tasks) in every job description and give a booster shot of training every day. You can utilize your existing computer systems to provide a brief reminder of these core critical tasks. One such system, as you know, is solid, realistic, ongoing, and verifiable training (SROVT). This training system is a process where six minutes is spent at the start of the workday with an emphasis on core critical tasks specific to the employee's job description. The goal is pre-incident knowledge level verification so things end up going right. Once again, if I were the lawyer representing your employee (my client) in a discipline matter, my second line of attack would be proof of when my client was trained on this policy. My follow up attack would be proof of when my client last received training on this policy? If you can't prove that my client was trained and retrained on it, your case will fall apart.

STRATEGIC HINT FOR YOUR CONSIDERATION

There must be consistent enforcement of organizational policy, and that begins by promoting people who have the ability to enforce policy and who understand supervision isn't a popularity contest.

You can't assume new employees will know how to perform an activity. Forget the faulty belief of common sense, as there's no such thing. New policies and procedures must be taught to employees regardless of how long they've been with your organization. You may find it helpful to introduce new policies as well as changes to existing policies at the same time (annually) so you can set aside a period of time for the necessary training of your personnel. Please take it seriously. Rule number one: Why do we need discipline? Rule number two: Build those good policies. And rule number three: Every day is a training day.

GRAED© #4

Make Sure Supervisors Are Behaving Like Supervisors

The key role of the supervisor is enforcement of organizational policy. Executives build rules (systems) and keep them up to date. Managers implement the rules, and supervisors enforce the rules. Not some of the rules, but all the rules. Not some of the time, but all of the time. Not with some of the people, but with all the people. Not one or two supervisors, but all supervisors. There must be consistent and across-the-board enforcement of organizational policy, and that begins by promoting people who have the ability to enforce policy. Promotional processes must filter out those who are unwilling or unable to enforce policy. Supervision isn't a popularity contest. Some personnel won't like you simply because you're enforcing organizational policy. Enforcement of policy must be consistent. When each supervisor enforces rules differently, you're setting multiple standards. Similarly, when rules are applied differently among employees, you're en route to problems. Supervisors must behave like supervisors.

Let me accentuate the point with a personal story. My father congratulated me when I promoted to sergeant in the California Highway Patrol (CHP) and imparted this sober advice: "Remember, last week, all you needed to do was follow policy. Now you're a supervisor. You need to enforce policy." And he followed up with this thought: "Gordon, some people are going to dislike you for enforcing policy." You know what I often hear from people when I tell them that story? "Gordon you're lucky your father was a cop." My father was never a cop. "Well, Gordon what'd your father do?" His big job was a chief engineer in the United States Merchant Marine during World War II, driving bombs from the west coast to the Philippines for the war effort.

For those of you who are military historians, one out of forty-eight US Marines died in World War II. One out of twenty-five US Merchant Marines died in World War II. Did you know that? The Japanese and the Germans were smart. They knew if they could break the supply chains, they could win the war. The Japanese had submarines off our west coast, and the Germans had submarines off our east coast. They knew if they could sink these cargo ships supplying England and the US Air Force and Army Air Corps in the Philippines, then they could win the war. So the United States had all sorts of systems to make sure their merchant marine vessels would be able to get through the enemy submarines. They had convoy, zigzag, and smokescreen operations. For boats loaded with bombs, they had a special rule: no smoking.

Now let's walk down this logical road together. Is there a reason you'd institute a no smoking policy on a boat loaded with bombs? Think it through. If you have enough seamen throwing matches and butts off the side of a boat, sooner or later one will find the boat or cargo and lose the bombs. You'll lose the crew, the ship, and the war. So on boats loaded with bombs, they had a no smoking policy. Just a quick thought for you. What percentage of men in World War II were smokers? I'll give you a hint. It was nearly a three-digit number. My dad was a smoker. In order for him to properly enforce the policy, he had to do what? Quit smoking. For some people, they had difficulty quitting. And when they wanted to smoke really bad, they went into nicotine rage. My dad had to enforce the no smoking policy.

You think you have a tough job as a supervisor. Oh, people might not like me. My dad was physically assaulted by merchant seamen who were under nicotine rage because he wouldn't let them smoke in the engine room. That's a tough job for a supervisor. Your job is to enforce organizational policy. And one more time. If I'm going to attack you on behalf of my client, your employee, I will proceed as follows: Show me the policy. "Here it is, Mr. Graham." Show me my client was trained on it. "Here's where he was trained on it, Mr. Graham." My next wave of attacks: Have other employees in your organization made the same mistake, and, if so, what happened to them? Were they similarly disciplined? If there's a lack of consistency, that could be the end of your discipline case. You must be consistent. When people don't follow rules, it must be addressed. So let's say you're disciplining an employee for not wearing a seatbelt. If I'm that lawyer, have other employees not worn seatbelts, or are you picking on my client? Does this have anything to do with my client's background? My client's race? My client's orientation? You understand where I'm going with this line of questioning. Lawyers are clever in their attacks. You must have a consistent across-the-board enforcement of organizational policy, and you must have supervisors behaving like supervisors.

STRATEGIC HINT FOR YOUR CONSIDERATION

Discipline must be initiated when there's deviation from policy. Any supervisor or manager who's aware of inappropriate, unsafe, illegal, or otherwise wrongful actions by an employee must act on this knowledge.

GRAED© #5

When Rules (Systems) Aren't Followed, Someone Needs to Act

Discipline needs to be initiated whenever there's an observed or discovered deviation from organizational policy. And that takes me back to my dad's comment to me on being promoted to sergeant: "Remember: Last week, all you needed to do was follow policy. Now you're a supervisor. You need to enforce policy." And he followed up with this thought: "Gordon, some people are going to dislike you for enforcing policy." What happens when people don't follow rules? When rules aren't being followed, somebody needs to act. Too many people want to take action only when there's a consequence (tragedy). That's not the issue. When rules aren't being followed, somebody must act. Discipline must be initiated when there's an observed or discovered deviation from policy. Any supervisor or manager who's aware of inappropriate, unsafe, illegal, or otherwise wrongful actions by an employee must act on this knowledge. Generally speaking, managers who become aware of deviations from established policy should inform the employee's immediate supervisor so she can take the necessary action. However, there are going to be some circumstances where the conduct is so egregious that a manager or executive must step in, particularly when it's a safety violation that could lead to serious consequences. Examples would include people not wearing their personal protective equipment (PPE) or not properly following safety protocols.

Discipline isn't dependent on consequences (tragedies). Waiting for a consequence such as a complaint, lawsuit, personnel injury, organizational embarrassment, internal investigation, or criminal filing isn't appropriate. Even if the underlying incident ends up going okay, any deviation from policy must be addressed. Failure by supervisors and managers to take action when they're aware of wrongful behavior ratifies the wrongful behavior and encourages future wrongful behavior. If you have sufficient future wrongful behaviors, sooner or later all the holes in the Swiss cheese will line up. The result is a consequence (tragedy), which must not be confused with proximate cause or the triggering event immediately preceding a tragedy. The real or root cause of many tragedies is a supervisor not behaving like a supervisor, or a supervisor trying to behave like a supervisor but not receiving management support. And then people will be sitting around saying what? That's been going on for years, and nobody did anything about.

STRATEGIC HINT FOR YOUR CONSIDERATION

Failure by supervisors and managers to take action when they're aware of wrongful behavior ratifies the wrongful behavior and encourages future wrongful behavior. That's a problem lying in wait.

The next chapter will examine the pervasive and explosive issue of job-based harassment. This type of misconduct must have a zero tolerance rule, and it must be enforced by supervisors. As a lawyer, I've seen the following script play out innumerable times: They've been telling jokes like that for years. They've been doing things like that for years. The supervisors were aware of it and nobody did anything about it. And then one day somebody gets angry and files a complaint. The big investigation is initiated by lawyers who confirm the preceding behaviors had been ratified for years. That's the classic behavior of Czar 52 and Lieutenant Colonel Bud Holland. Everybody knew he had all these bad behaviors but nobody did anything about it because there were no consequences (tragedies) from his behavior. In other words, all's well that ends well. The only time his superiors addressed the problem was post-incident after tragedy struck. Don't wait for the tragedies. Act up front.

GRAED© #6

Take the Time to Think It Through, and Keep It Private

Your job as a supervisor or manager is a series of activities. Some of these activities require immediate action (non-discretionary time), whereas others allow you time to think (discretionary time) to make a decision. As you move up the ranks, most of your activities involve time to think. Employment law is one of those discretionary time functions. Take the time to think it through and slow down.

You have considerable time to make the decision. Once the decision has been made, it is difficult to un-ring the bell without embarrassment or other nasty complications.

STRATEGIC HINT FOR YOUR CONSIDERATION

Slow down and take the time to properly think through your disciplinary decision. Once the decision has been made, it's difficult to un-ring the bell without embarrassment or other nasty complications.

Gather and review all of the facts. "I know what happened there." I've had so many cases where an employee did something, and her supervisor prematurely made up her mind. "Well, he said this, this, and this. That's a violation of policy." Maybe there's a reason it justified the behavior. "Well, he touched her inappropriately." Maybe he was pulling her away from an obvious risk that she didn't see? Please don't allow yourself to suffer cognitive lock. When we talk about decision making in Chapter 10, I'll introduce this concept in greater detail. Briefly, cognitive lock occurs when the brain makes up its mind on partial information. When we get locked in a course of behavior, we can ignore exculpatory evidence. "I don't like this employee. I saw the employee do that. Therefore the employee is guilty." Talk to the employee in advance, and remember the importance of consistency.

If applicable, it's imperative you review the MOU (memorandum of understanding) your organization may have with the involved employee's representing unit. In some jobs, there may even be laws that impact the discipline process. You can further use this discretionary time to get in touch with your human resources personnel, as they'll be more familiar with the idiosyncrasies of discipline than you. If the activity is severe, you should consult with competent counsel who'll defend you and your organization downstream. Getting their input up front can prevent subsequent problems. The general rule is to have substantial evidence of misconduct in order to substantiate discipline. One of the rules that many good supervisors follow is praise in public but discipline in private. That's a good rule to follow, and it doesn't mean you allow unsafe or inappropriate behaviors to continue simply because other employees are present. It means you take necessary action to stop the inappropriate conduct, and you take appropriate action against the employee in private after you've gathered your facts, talked to your manager, human resources personnel, or your competent counsel. Gather all your facts. Call him in and be professional. Explain the violation of policy and what you're going to do about it. Good people, and I believe you have them, will welcome prompt, fair, and impartial discipline. What they won't accept is delayed, unfair, and partial discipline. Please take it seriously.

GRAED© #7

Be Specific When Talking to the Involved Employee

STRATEGIC HINT FOR YOUR CONSIDERATION

Unfair discipline will cause substantive morale issues, which will impact the organization more severely than the underlying employee misconduct. Employees expect and deserve prompt, fair, and impartial discipline.

STRATEGIC HINT FOR YOUR CONSIDERATION

During the discipline process, you must be specific. This specificity isn't only to the respective behavior behind the discipline but also to the facts surrounding the activity and the actions you're taking.

During the discipline process, you must be specific. This specificity isn't only to the respective behavior behind the discipline but also to the facts surrounding the activity and the actions you're taking. Present the policy, rule, or procedure violated, and clearly state the specifics of the violation. You should talk to the employee in a professional manner and not allow his comments or behavior to change your demeanor. Attack the behavior; don't attack the employee. Insubordination can and should be addressed if it's present. Don't use this opportunity to bring up past problems, unless they're directly related to the current problem. As discussed in rule number four, this interview should be documented so there's no doubt downstream as to what transpired. You should also have another supervisor as a witness. Remember if it isn't documented, it never happened. And going back to Czar 52, all the superiors at Fairchild Air Force Base talked to Bud Holland about his reckless behavior but nobody wrote anything down. High-reliability organizations intuitively understand discipline is a conditioned human behavior that requires ongoing training and enterprise commitment.

Let me share an example of specificity. There's a new rule on airplanes. If you check a bag, you must remove any lithium batteries. I've lithium batteries in my earphones, cell phone, and iPad. I carry these items on the airplane. What's the big deal? So I finally asked. If you have lithium batteries in your checked luggage, why do you need to take them out? I can carry them onto the cabin so why can't they be checked in my luggage? "Mr. Graham, if they ignite in the cabin, we're aware of it. We can put the fire out. If they ignite in the cargo hold, we might not be aware of it until the fire becomes substantial. At that point, the aircraft may be compromised, and our passengers may be in grave danger." Specificity. Why do we remove the lithium batteries? Please be specific.

GRAED© #8

Keep It Professional, Fair, and Balanced

For organizational discipline to achieve its intended goal of preventing future problems, it must be professional, fair, and balanced. Did it ever trouble you as a former line-personnel that certain people avoided discipline more than others? Did some employees get away with a lot, while others were badgered for the most minor of infractions? Is the degree of discipline you plan to administer reasonably relative to the seriousness of the proven offense? Is it fair considering the record of the employee's service with your organization? Is it consistent with the discipline that's been meted out to other employees involved in the same or similar behavior? Unfair discipline will cause substantive morale issues, which will impact the organization more severely than the underlying employee misconduct. Good employees welcome prompt, fair, and impartial discipline. What they don't like is delayed, unfair, and partial discipline. If I were the lawyer representing an employee you're disciplining, I'd have several different types of attacks: Show me the policy. "Here it is." Show me they were trained on it. "Here it is." Show me the supervisors were enforcing the policy. "Here it is." And my final attack would be: Have other people been disciplined differently for this? And if your answer is yes, that inconsistency could be a big problem for you.

STRATEGIC HINT FOR YOUR CONSIDERATION

Your documentation today will establish the record, justify the imposed action, and serve as a defense if downstream allegations are made against you for taking the action. It's absolutely essential.

GRAED© #9

Documentation Is Essential

Meting out discipline when there's an observed deviation from organizational policy is a required activity for a supervisor or manager. It's a serious matter because the disciplinary action may result in loss of pay, time off, negative performance review, and other undesirable actions for an employee. As such, the entire process must be fully documented. Also, the rules of progressive discipline for subsequent deviations from organizational policy require full documentation. If it isn't written down, many will believe it didn't happen. Always remember the value of documentation through the feedback loop of managing by walking around (MBWA) and supervising by walking around (SBWA). At a minimum, don't forget the 5Ws (who, what, when, where, and why) and the 2Hs (how many and how much) in your information and problem-solving phase. Your documentation today will establish the record, justify the imposed action, and serve as a defense if downstream allegations are made against you for taking the action. Take the time to document exactly why you did, what you did, when you did it. If it isn't written down, it didn't happen.

GRAED© #10

Get on with Life and Your Organization's Mission

Once the process is over, both the supervisor and employee must get back to their respective responsibilities. Internalizing the issue by either party will cause future problems and adversely impact organizational dynamics. If the inappropriate behavior occurs again, it needs to be addressed again and again as necessary. However, constantly bringing up past disciplinary problems that have been concluded is wrongheaded. Get on with life and the mission of your organization. You've taken your discipline, or you've received your discipline. Sadly, I meet people who never forget the violation, and they're always reiterating it. That's not healthy or productive. Remember, your role as a supervisor is to enforce organizational policy and to facilitate the performance of your subordinates. When you perform your role correctly, everybody wins, both inside and outside of your organization.

FINAL THOUGHTS

Organizational risk management is the crux of *real* risk management. I believe you have good people. I know you can get good policy. You must make every day a training day and focus it on core critical tasks and critical tasks. The catalyst of high-reliability organizations is its supervisors. To be successful, your supervisors must behave like supervisors. Their primary mission is enforcing organizational policy. On those rare occasions when people don't follow rules, it must be addressed with prompt, fair, and impartial discipline that isn't a function of consequence but a function of policy. Without discipline, the other pillars of success become ineffective and your organization will spiral down to mediocrity. At that level, there's an overrepresentation of tragedies and underrepresentation of employee fitness and customer loyalty.

We have two chapters left, with Chapter 9 dealing with job-based harassment and Chapter 10 focusing on my favorite topic: ethical and sound decision making. Until then, please work safely.

Chapter Takeaway on Media Relations/Crisis Communication

Please refer to the media relations/crisis communication material in the Addendum section and ask yourself how you would address the following scenario:

Peter was one of your most talented distribution operators. His skill was unmatched, as was his beyond-life personality. The problem was Peter didn't follow the rules. He knew he was good and was arrogant. His supervisor also realized it and always warned him to follow policy. Unfortunately, there was no documentation and no discipline when he violated policy. Everyone in the organization knew Peter was a talented but rogue employee. They also saw how the rules didn't apply to him. Peter was treated differently, and that upset the other operators and impacted morale.

Today, Peter was killed in a trench that he chose not to properly shore. There were two other employees in the trench when it collapsed. Those employees are fighting for their lives and will have permanent brain injuries if they pull through. You've never lost an employee to an on-the-job fatality, let alone a tragedy involving two life-threatening injuries. You're grief stricken and scared. There was a clear failure in leadership, and you lead the organization. Is it systemic? Will these problems lying in wait surface from the OSHA investigation? What about the workers' compensation premium increases and likely policy nonrenewal? How will you explain those issues to your governing body? And what if another tragedy occurs by a rogue employee who wasn't disciplined for violating the rules. Can your beleaguered organization redeem employee confidence?

The local news and print media are aware of the tragedy and badgering you for an interview. The more you hold them off, the more likely they'll interview your employees. That concerns you. Will they say Peter was a rogue employee and the tragedy was preventable? Will they share their frustration that rules didn't apply to him? You have to schedule a press conference and provide answers to the community, your employees, and grieving family members. How much do you share with the public? Do you disclose these organizational failures? And how do you console the families without arming them with multi-million-dollar facts for an inevitable unsafe workplace lawsuit. What can you do to balance these competing objectives while still being honest, authentic, and grief stricken? How can you use this press conference to regain some semblance of control?

9 Job-Based Harassment

Zero Tolerance Is the Rule

2019 Photo Courtesy of California Water Service Group

SUMMARY

Job-based harassment is a pervasive condition that exists throughout the United States. This timeworn exposure isn't a black swan but a preventable gray rhino, and the consequences for allowing such debauched behavior are serious, including failure of organizational mission and plunging employee morale.

There are various legal jeopardies linked to this exposure, which necessitate a broad job-based harassment policy inclusive of background investigations, personnel training, zero tolerance, and analyses of actions against protected classes.

Chapter emphasis will center on the key role of supervisors in setting examples and enforcing policies as well as providing harassment victims a complete and impartial investigation where zero tolerance and eradication of all inappropriate behavior are the organizational objectives.

DOI: 10.1201/9781003229087-9

Hello again. Gordon Graham here and welcome back to our pursuit of *real* risk management. As promised, we'll be pivoting our final two chapters from a singular discussion on the five pillars of success to job-based harassment and ethical decision making. The former, *job-based harassment*, is pervasive in all professions, and such inappropriate and predatory conduct is timeworn. The Me-Too movement and its innumerable cases of debauchery from public figures have manifested a cause célèbre where societal and legal tolerance no longer condones such behavior. Job-based harassment in our great country has unfortunately been around since its founding. The same applies to racial, religious, and gender discrimination as well as lesbian, gay, bisexual, transgender, and queer (LGBTQ) harassment. This conventional thinking began to shift in the 1940s.

The actual line of demarcation was World War II. Before the war, there were specific jobs for African-Americans, Hispanic-Americans, Asian-Americans, and other racial groups. Women were also negatively impacted, as was the LGBTQ community. And then the war happened, and we needed one-hundred percent of our population to fight the common enemy. All of a sudden, African-Americans were performing nontraditional jobs as were Mexican-Americans, Asian-Americans, and women. Everybody worked together, and the United States won the war. After the war, society tried to restore the status quo. African-Americans had a complaint: "I flew these airplanes in World War II. How come I can't fly them now as an airline pilot?" Mexican-Americans had a complaint: "I was a contractor in World War II building those massive structures for the military. How come I can't be a contractor now?" Women had a complaint. "Look what I did during World War II? I assembled those planes and tanks. How come I can't have a manufacturing job now?" Society tried to restore the original order, and it didn't work. The result was civil unrest and a national progression toward equality.

It all came to a head in the early 1960s when John Kennedy was elected. Contrary to what you might've thought, President Kennedy was initially opposed to federal intervention in civil rights issues. He thought the states could take care of it. Remember, Kennedy was a US senator. Quickly into his presidency, he realized many states weren't taking civil rights seriously and federal intervention would be necessary. Before his assassination in 1963, he began working on the Federal Civil Rights Act, which ultimately passed in 1964. This law created Title VII and an enforcement mechanism called the Equal Employment Opportunity Commission (EEOC).

Almost every harassment, bias, and discrimination law in every state started in the early 1970s. People, however, were still being treated inappropriately in various workplaces. Harassment, bias, and discrimination based on protected class status of race, sex, and religion were rampant. As a result, the courts started hearing these cases in the 1970s through the enforcement mechanism of Title VII, the EEOC. Figure 9.1 illustrates the various types of legal jeopardy that employers encounter from job-based harassment. It also identifies the various options available to employees who are subjected to inappropriate behavior.

The EEOC is a domain of lawyers. Below the EEOC is the state human rights commission. That's also the domain of lawyers. Next to that are civil lawsuits. Again, the domain of lawyers. We then have criminal investigations if the conduct is egregious enough. The domain of lawyers. And finally, the internal investigation. The domain

Employer Exposure to Job Based Harassment
Types of Jeopardy
➲ Equal Employment Opportunity Commission (EEOC)
➲ State Human Rights Commission (SHRC)
➲ Civil
➲ Criminal
➲ Internal

FIGURE 9.1 Employer exposure to job-based harassment. Source: Graham Research Consultants, LLC.

of lawyers. Those are the consequences (tragedies) when people aren't being treated fairly at work. Employees have the federal option, state option, civil option, and internal option. The criminal option is reserved for local, state, and federal prosecutors. I'm jumping several pages, but of the five, which one do you hope your aggrieved employee chooses? If you're being honest, you want him to choose number five: the internal option. It's the quickest, cheapest, and easiest way to resolve these issues. Let me share a hint: If people don't have faith in the internal process, they'll abandon this option and go external. That's when you have major problems. More on that in just a bit.

STRATEGIC HINT FOR YOUR CONSIDERATION

Most victims of harassment, whether it be sexual, racial, religious, sexual orientation, or any other protected class, don't want lawsuits or external investigations but rather for the inappropriate behavior to stop.

The courts, and they're all lawyers, were looking at the behaviors that generated allegations of bias, harassment, and discrimination. There were thousands of specific behaviors that produced these complaints, including inappropriate conduct, inappropriate comments, inappropriate actions, and inappropriate things going on in the workplace. The courts compiled this assortment of inappropriateness and cataloged them into three families: quid pro quo; retaliation; and hostile work environment. It developed and evolved from there. This registry of inappropriateness within three families occurred in the 1970s. Fifty years later, it's still three families. You read the advance sheets released from the courts around our great country, and the allegations remain compartmentalized under these three families as illustrated in Figure 9.2: quid pro quo; retaliation; and hostile work environment.

Let's take these one at a time so you can identify what it means and what you can do up front as a risk manager to prevent these things from happening. Quid pro quo is a Latin expression. Raw translation: this for that. You do this for me, and I'll do that for you. There's nothing wrong with quid pro quo. My relationship with my clients is quid pro quo. Do this work, and we'll pay you. I have a quid pro quo relationship with my wife. I take her to cop parties, and she takes me to dog parties. That's quid

Types of Job Based Harassment / Legal Triggers

- Quid Pro Quo
- Retaliation
- Hostile Work Environment

FIGURE 9.2 Types of job-based harassment/legal triggers. Source: Graham Research Consultants, LLC.

pro quo. You do this for me, and I'll do that for you. I have a quid pro quo relationship with my daughter. You get good grades, and I'll pay your tuition. Not all examples of quid pro quo are proper. Have sex with me, or you're fired. Have sex with me, and I'll get you a new car. Have sex with me, and I'll get you that overtime. Have sex with me, and I'll get you a better assignment. That's illegal, immoral, and a big deal. Guess what? That happens in organizations throughout our great country. There are literally thousands of specific behaviors that generate quid pro quo allegations.

Let's go beyond the proximate cause and look for root cause. There are two root causational factors leading to quid pro quo in the workplace. Number one: Don't hire or promote losers. Utilize best practices as it relates to recruitment, backgrounds, probation, and performance evaluations. The best predictor of future behavior is past behavior. Leopards don't change their spots, and you can't train the immoral to be moral. Perform your background checks, and don't hire losers. Number two: Don't attempt to date people who work for you. Nothing but bad will happen. Even if it's legal in your state, don't do it. Let's accentuate the point with an example.

I'm an executive for a water-related entity, and I asked you out. You agree, and we're in love. Now, I have to pick somebody to inspect a new water purification technology from Switzerland. I evaluate all my employees, and the person I love is the most qualified. Even if this employee is the most qualified, there's a different perception from others in the workplace. That perception is reality. Here's another reality for you. I asked you out, and you said no. No harm and no foul. I asked you out, you declined, and we resumed our respective duties. In reality, have I given you a golden defense for any employer action filed against you? Five years from now, I'm trying to fire you for theft. I've given your lawyer a guaranteed defense: "Excuse me, executive Graham. Did you or did you not ask out my client five years ago?" I did. "And what did she say." No. You know what that says to the jury? "I'm getting even with this person for rejecting my sexual advances." Don't hire or promote losers, and don't date people that work for you. Nothing but bad will happen. That's quid pro quo.

The second one is retaliation. What's retaliation? That's when someone files a complaint and people gang up and treat him inappropriately because he filed a complaint. It's illegal, immoral, and a big deal. It can also extend to other employees who are friends and family members of the complainant who work for the same organization. There are several proximate causes for that allegation but only two root causational factors. Number one: Don't hire or promote losers, and always perform comprehensive background investigations. Number two: performance evaluations. Go ahead. Keep overrating people because it's easy. "You're doing great." Everything's wonderful, and then the employee files a complaint against your

Exposure Drilldown

- Quid Pro Quo
  - Hiring
  - Dating
- Retaliation
  - Hiring
  - Performance Evaluation

Question: What's the common denominator for quid pro quo and retaliation?

Answer: Supervisors not behaving like supervisors; or supervisors behaving like supervisors but not receiving management support.

FIGURE 9.3 Exposure drilldown. Source: Graham Research Consultants, LLC.

organization and names you as a defendant. You're now angry. Did he get an overrated performance evaluation this year? No. Needs improvement. Needs improvement. Needs improvement. Doesn't meet standards. That sounds like retaliation to a jury. He was doing fine until he filed a complaint, but, in reality, he was never doing fine. We overrate people because it's easy, and those documents can haunt us. Figure 9.3 illustrates the root causational factors and common denominator of quid pro quo and retaliation.

STRATEGIC HINT FOR YOUR CONSIDERATION

There are thousands of specific behaviors that produce job-based harassment complaints, and the courts have cataloged this inappropriateness into three families: quid pro quo; retaliation; and hostile work environment.

Quid pro quo and retaliation, however, aren't the Goliaths of job-based harassment. Most of the complaints filed against water-related entities aren't quid pro quo or retaliation. They're hostile work environment. What's the definition of a hostile work environment? Severe, pervasive, pattern, and practice of misconduct directed to a protected class status. There are two root causational factors for a hostile work environment. Number one: supervisors not behaving like supervisors. I've exhausted my warnings on that topic. The primary mission of a supervisor is enforcement of organizational policy. If you have supervisors who aren't enforcing the policy, that's a problem lying in wait. Number two: supervisors who try to behave like supervisors but receive no support from management. Both are problems lying in wait. That's the three families of harassment.

What can you do to address job-based harassment? I have a four-hour class on harassment. Here's the way I start my harassment class. Good morning ladies and gentlemen. Gordon Graham here. I have you for four hours on job-based harassment. Let's get started. And I reach into my briefcase and pull out a stack of index cards. I then walk through the group, and I give the index cards to every woman in the group. You can hear people talking while I'm doing that. "What's he doing? He's only giving the cards to the women." I return to the front and say the following: Ladies, would you please look around the room and write down your organization's five biggest sexual predators? Then, I pick up the cards. I've been performing this routine for years. In every stack of cards, there's always one name mentioned with higher frequency. I hold the cards up to the police or fire chief and say: Chief, I'll sell you these cards for twenty-five-thousand dollars, and you can see her thinking. Obviously, I'm joking and rip them up. The reality is you don't need to buy them. You likely already know the problem offenders. Regrettably, these bad actors are problems lying in wait. They can and will cause serious consequences (tragedies) to your organization.

Once again, it's a matter of will. If you take that employee on to stop the inappropriate behavior before a complaint is filed, you get paid X. If you don't do anything, you get paid X. Mediocrity replaces accountability. We must take this issue seriously. Get a good policy and enforce the policy: people, policy, training, supervision, and discipline. Harassment in the workplace isn't a black swan but rather a gray rhino, and the consequences for allowing inappropriate behavior in the workplace are significant. High-reliability organizations implement a zero tolerance policy for job-based harassment because it interferes with their inexorable pursuit of doing the right thing, the right way every time. The lack of a zero tolerance policy will lead to mission failure and organizational tragedy as well as a vacuum in employee fitness and customer loyalty.

What can you do up front to address job-based harassment? You must get yourself a good policy that's regularly reviewed by competent counsel as well as subject matter experts. This policy must be compliant with state and federal law. It must similarly include a harassment policy that covers all protected classes. If you haven't updated your harassment policy in the last year, it's probably out of date. Why is that? Employment law is continually changing and evolving through legal precedent and regulatory requirements at both the state and federal levels. There are ten activity-specific rules for eliminating sexual harassment, which I call GRESH© (Graham's rules for eliminating sexual harassment). All of the versions of harassment are included in these rules, and they're based on the five pillars of success. These rules involve getting and keeping good people as well as getting and obtaining good policies. The other pillars of training, supervision, and discipline are also prominently interlinked. Figure 9.4 illustrates the components of GRESH©. Let's examine each of these rules in detail.

GRESH© #1

Ensure You Have a Comprehensive Policy (System) That's Organizational Specific, Properly Derived, and Regularly Reviewed by Your Competent Counsel to Be in Compliance with This Rapidly Changing Area of State and Federal Law

Zero tolerance must be the rule. Make sure your policy (system) includes a section on how to report and how to investigate deviations from established policy. If your

Graham's Rules for Eliminating Sexual Harassment (GRESH©)		
GRESH© #1	➲	Ensure you have a comprehensive policy (system) that's organizational specific, properly derived, and regularly reviewed by your competent counsel to be in compliance with this rapidly changing area of state and federal law. Zero tolerance must be the rule.
GRESH© #2	➲	Don't hire losers!
GRESH© #3	➲	Make sure your employees have a one-hundred percent knowledge level verification of the organization's harassment policy.
GRESH© #4	➲	Set the proper example.
GRESH© #5	➲	Remember the principles of risk management, and use the CATSINRO© analysis on a regular basis: Comments, Actions and Things, that are Sexual, Racial, Religious, or Sexual Orientation in nature in which Reasonable People would find Offensive. These activities don't belong at work and can't be tolerated.
GRESH© #6	➲	Discourage supervisors and managers from dating subordinates.
GRESH© #7	➲	Ensure you have a system for the prompt, fair, and impartial investigation of all complaints.
GRESH© #8	➲	Forget the welcome versus unwelcome analysis. It's unworkable. Rather, ask if a reasonable person would find the behavior inappropriate.
GRESH© #9	➲	There's a definite link between the way your organization treats its employees, the way your employees treat each other, and the way your employees treat customers.
GRESH© #10	➲	Most things that go wrong in life are highly predictable, and predictable is preventable.

FIGURE 9.4 Graham's rules for eliminating sexual harassment (GRESH©). Source: Graham Research Consultants, LLC.

policy hasn't been updated to reflect any decisions from the most recent term of the US Supreme Court as well as new state and federal regulatory requirements, then it may be out of date. If your policy addresses sex only, it's definitely out of date. To adequately address job-based harassment, you must get yourself a good policy and encase it with the other four pillars of people, training, supervision, and discipline.

GRESH© #2

Don't Hire Losers

Build your organization's commitment for eliminating harassment into the hiring process. Job announcements must contain this commitment. During the hiring process, ask specific questions requiring specific answers as much as the law allows prior to employment. Don't hire those who are unwilling or unable to abide with your organization's policy. Remember, the best predictor of future behavior is past behavior. Make sure you perform comprehensive background investigations and take the process seriously. A thorough investigation will filter out losers and ensure your organization is represented by good people and not bad actors.

GRESH© #3

Make Sure Your Employees Have a One-Hundred Percent Knowledge Level Verification of Your Organization's Harassment Policy

Knowledge is more than a document certifying you've reviewed the policy or attended a training session. Knowledge means you're regularly (at least quarterly) covering the policy and making sure each employee knows what it says. Harassment isn't a core critical task, but it's one of those high-risk areas that must be covered at least annually and requires high-risk training. Personnel can't follow policy if they don't know policy.

STRATEGIC HINT FOR YOUR CONSIDERATION

An excellent time to discuss job-based harassment is during the annual performance evaluation process. Make sure your people understand the organization's commitment to eliminating harassment, bias, and discrimination.

STRATEGIC HINT FOR YOUR CONSIDERATION

Zero tolerance must be the rule. Make sure your job-based harassment policy is updated annually and includes a section on how to report and how to investigate deviations from established policy.

They won't know policy if the last time they read it was during a training class some time ago. You have posters in your facility to reemphasize the importance of occupational safety. Is harassment less important? An excellent time to talk about harassment is once a year during the performance evaluation process to make sure people understand your organization's commitment to eliminating harassment, bias, and discrimination as well as the enforcement mechanisms when such behavior surfaces.

GRESH© #4

Set the Proper Example

Personnel won't follow policy if it isn't followed by management and supervisory personnel. Managers and supervisors who are aware of inappropriate behavior and fail to enforce the policy have ratified the behavior. And that encourages future inappropriate behavior. The primary mission of a supervisor is enforcement of policy. Build these convictions into your internal promotional process. Don't promote losers. Run a tight ship. Promote people who'll enforce the policy. For those of you who are executives, managers, and supervisors, set the proper example.

GRESH© #5

Remember the Principles of Risk Management

The stakes are too high to wait until an incident occurs. Remember the three precepts of *real* risk management: (1) There's no new ways to get in trouble; (2) there's always a better way to stay out of trouble; and (3) predictable is preventable. If you can identify the risk, you can put together control measures (systems) up front. The control measure I put together to address harassment in the workplace is called CATSINRO©.

When I started studying quid pro quo, retaliation, and hostile work environment allegations, I came up with three behaviors that were overrepresented in downstream litigation: inappropriate comments, actions, and things at work. And out of that came CATSINRO©. Before you say something, before you do something, before you bring something to work, you must analyze what you're doing and apply the CATSINRO© analysis. It stands for Comments, Actions, and Things that are Sexual, Racial, Religious, or Sexual Orientation in nature which Reasonable People would find Offensive. Such behavior doesn't belong at work. You'll never have a problem downstream if you eliminate this behavior. Figure 9.5 illustrates CATSINRO©.

Managers and supervisors must regularly inspect the workplace and look for CATSINRO©. That's what managing by walking around (MBWA) and supervising by walking around (SBWA) are all about. Walk around your workplace and look for problems. Walk into the locker room. If there are inappropriate posters in the locker room, that's a problem lying in wait. Listen to what people are saying. If there are inappropriate jokes being shared in the workplace, that's a problem lying in wait. Listen to what people are saying about the way they're treated. If there's potential misconduct, that's a problem lying in wait. Additionally, use the APE (annual performance evaluation) process to review the policy with all personnel, and look for minor issues prior to their becoming major issues. Ask the APE questions: Are you aware of our organization's commitment? Are you aware of any inappropriate behavior occurring in the workplace? Ask those questions on a regular basis. Such

➲	Comments, Actions and Things, Sexual, Racial, Religious or Sexual Orientation in nature which Reasonable People would find Offensive don't belong at work.
➲	Eliminate the behaviors upfront and you'll never have a problem downstream.
➲	Zero tolerance is the rule.

FIGURE 9.5 CATSINRO©. Source: Graham Research Consultants, LLC.

an action sends a serious message to all employees. Good women and men will rise and fall to your level of expectation, so set the standards high.

GRESH© #6

Discourage Managers and Supervisors from Dating Subordinates

Dating subordinates is inane. This situation is ripe for quid pro quo allegations and sets the wrong tone for the rest of the workplace. For many observers, perception is reality. What perceptions are created when managers and supervisors are dating subordinates? As much as legally possible, discourage dating between supervisors and employees. Only bad things will happen.

GRESH© #7

Ensure You Have a Program for the Prompt, Fair, and Impartial Investigation of All Complaints

Pick investigators who'll represent your organization in an appropriate manner. Not every manager or supervisor is qualified to be an investigator. Make sure you have a protocol for the standardization of job-based harassment investigations. If you're hiring outside people to perform these investigations, make sure the process is known and taken seriously. Most victims of harassment, whether it be sexual, racial, religious, sexual orientation, or any other protected class, don't want lawsuits or external investigations. They want the inappropriate behavior to stop. If they don't believe the investigation will be fair, then they'll use the external remedies. Ensure you have a system for the prompt, fair, and impartial investigation of all complaints. Please take all job-based harassment complaints seriously and make sure your employees understand your commitment to a zero tolerance policy.

STRATEGIC HINT FOR YOUR CONSIDERATION

Dating subordinates is ripe for quid pro quo allegations and sets the wrong tone for the rest of the workplace. Always remember that perception is reality, and, as much as legally possible, discourage this behavior.

GRESH© #8

Forget the Welcome versus Unwelcome Analysis

From a risk management standpoint, the welcome versus unwelcome standard is unworkable. Rather, ask if a reasonable person would find the behavior inappropriate.

If so, remove the behavior and take appropriate discipline. What appears to be welcome behavior today can be deemed unwelcome behavior tomorrow. How do you know the behavior was welcome? Zero tolerance must be the rule. I'm not normally an advocate of zero tolerance, but I don't know how else to say it. Do your CATSINRO© analysis: Comments, Actions, and Things that are Sexual, Racial, Religious, or Sexual Orientation in nature which Reasonable People would find Offensive don't belong at work.

GRESH© #9

There's a Definite Link between the Way Your Organization Treats Its Employees, the Way Your Employees Treat Each Other, and the Way Your Employees Treat Customers

You can't fix the outside until the inside is running smoothly. There's a link between the way your organization treats its employees, the way employees treat each other, and the way your employees treat customers. You show me an organization that's plagued internally with inappropriate behaviors, and I'll show you the same behaviors externally. Alternatively, you show me an organization where the rule is dignity and respect internally, you'll see the same behavior manifested externally. Customer loyalty, community goodwill, and employee morale are only achieved with a zero tolerance policy on job-based harassment.

STRATEGIC HINT FOR YOUR CONSIDERATION

There's a link between the way your organization treats its employees, the way employees treat each other, and the way employees treat customers. A zero tolerance job-based harassment policy creates customer loyalty.

GRESH© #10

Most Things That Go Wrong in Life Are Highly Predictable

That's also true with job-based harassment. Establish the proper culture. Recognize the five pillars of a good organization: people, policy, training, supervision, and discipline. Take one of these away, and tragedy will ensue. Preventing problems is easy. Treat people the way you'd like to be treated. If you don't have something good to say, don't say anything. Keep your hands to yourself. And you attract more flies with sugar than with vinegar. The solution isn't complex and correlates to precept three: predictable is preventable. Things that go wrong in life are predictable, and predictable is preventable.

FINAL COMMENTS

The way you avoid problems downstream in Lawyerville is through the five pillars of success: people, policy, training, supervision, and discipline. Take one away and tragedy will ensue. Preventing job-based harassment is easy as long as you treat people the way you'd like to be treated. The solution isn't complex because the rule of zero tolerance, coupled with comprehensive and impartial investigations, removes all ambiguity. Those are my thoughts on job-based harassment in this penultimate chapter. We have one left, and it's on my favorite subject: ethical and sound decision making. Allow me to provide a primer: In many organizations, we teach our people how to do things, but we don't teach them how to think. I'll be imparting a sequential and logical checklist, or a ten-step process, on how to think things through. I look forward to seeing you. Until then, work safely.

Chapter Takeaway on Media Relations/Crisis Communication

Please refer to the media relations/crisis communication material in the Addendum section and ask yourself how you would address the following scenario:

You've been a proud member of your organization's leadership team for nearly fifteen years. You've seen its impact on the community, both in terms of philanthropic endeavors and direct outreach. Equally important, the organization prides itself on being a meritocracy where hard work and professional development are rewarded. Protected classes have always been treated fairly and properly. Over the last week, that entire predicate has been challenged. Nearly one-fifth of your line-personnel signed a letter of support for six of their colleagues who recently filed a hostile work environment suit for discrimination. In the complaint, the plaintiffs cite organizational leadership as tacitly condoning inequalities in promotion, pay, and scheduling. They also mention lack of diversity in leadership as proof of your organization's inclusionary facade.

You've never been accused of being a racist and have never seen inappropriate or unfair treatment toward any employee, including protected classes. Your mind quickly rotates through the concrete measures your organization implemented to promote inclusion. All the training courses, diversity memos, community support, and recruitment efforts are now on repeat cycle. These initiatives came at great expense to the organization, and your annoyance turns to anger. How dare these plaintiffs and the employees who signed that letter of support say I'm a racist or that our organization is racist. They can't name a single instance of outrageous or unacceptable behavior.

It then hits you. What about the inappropriate jokes, comments, and materials you observed throughout the years? You also recalled deleting several email chains because they weren't suitable for work and clearly insensitive. Those were just isolated events, right? It was also true our leadership team wasn't properly diverse, but that's only because of a lack of protected class applicants. No one else applied. And what about the disproportionality of diversity in our managerial and supervisory ranks? I know we're not matching our community, but it's not our fault. We're a meritocracy, and that means individuals are expected to take initiative to promote. I can't be wrong, can I? These memories of unacceptable workplace behavior are troubling.

Your organization is now being presented as evil, and that characterization is supported by a good percentage of your employees. The press has taken control of the narrative and cast you as the antagonist. They're contacting employees and citing anonymous sources with stories that are fantastical and stretch all credulity. Yet, they're reported as fact without any opportunity for the organization to clinically and fairly address. You know it's time to take back the messaging. How do you do so without exacerbating the internal turmoil? Do you own the problem so you can focus on the solution? Or will that make the situation worse and create a financial exposure with the ongoing litigation and motivate similar lawsuits?

Tomorrow morning, you'll address these issues with a formal press conference. You'll have prepared remarks for the litigation piece, but you must speak openly and compassionately on the overarching issue of discrimination in all workforces. There's opportunity in every crisis, and you must use this situation as a platform to confront this black swan event, which is actually a stampede of gray rhinos in disguise. It's evident your organization made mistakes. How do you admit failings while not accepting liability yet still offer a clear pathway for a better tomorrow based on real inclusion, real diversity, and real change to your organizational culture, practices, and expectations?

10 Ethical Decision Making

Thinking through the Consequences

STRATEGIC HINT FOR YOUR CONSIDERATION

Systems and culture play an integral role in achieving organizational mission, but its impetus is employees who've been taught how to think through their activities by applying an ethical and sound decision making process.

DOI: 10.1201/9781003229087-10

SUMMARY

Ethical and sound decision making is inculcated in high-reliability organizations where employees are taught to fully and deliberatively think through their discretionary time activities. This learned behavior ensures employees do the right thing, the right way every time, and also warrants they know why it's the right thing to do.

Ethical and sound decision making isn't inherently known by new employees and must be taught through initial and ongoing training to maintain consistent and accurate performance of activities that comport to organizational mission and expectations.

Chapter emphasis will focus on ten sequential rules predicated on two core questions: What's going on here, and what I'm being asked to do or decide? Answering those questions and following the next steps will ensure employees think through the consequences of their decisions.

Hello again. Gordon Graham here, and welcome back to our journey toward *real* risk management. We're approaching our final destination, with this chapter and a conclusion remaining. Chapter 10, "Ethical Decision Making," captures all aspects of my philosophy of doing the right thing, the right way every time. We know *real* risk management is embodied by high-reliability organizations and impelled through continuous improvement of resilient systems built from root causational factors. Resilient systems, however, aren't enough to prevent tragedies. These enterprises also implant a culture of discipline where employees understand, support, and enforce prescribed systems and their underlying rationale.

The interrelated components of systems and culture play an integral role in achieving organizational mission and purpose, but its impetus is employees who've been taught how to consistently, constantly, and accurately think through their activities by applying a deliberative, methodical, and ethical decision making process. This catalyst will be our emphasis and will encompass a process of activity-specific rules to achieve your overarching goal of doing the right thing, the right way every time while always keeping the consequences of one's decisions in the fore of deliberation.

I'm frequently invited to serve as a consultant for public safety organizations throughout our great country. As an advisor, I always try to attend my clients' staff meetings. Organizational dynamics are on full display in these meetings. That's where the executives gather around the table. Who sits where? Who commands respect? Who's taking notes? My guess is water-related entities also have staff meetings where executives congregate. Here's my experience in these meetings: The roundtable discussion is the last item on the agenda. All right folks, we've completed the agenda. It's now time for our roundtable discussion. Mary, how's everything in your department? "Pretty good, chief." Steve, how's everything in your department? "Pretty good, chief." Sara, how's everything in your department? "Pretty good, chief." Fred, what went wrong in your department? "Well, chief, the involved employee made a bad decision." Really? The involved employee made a bad decision? And I look around the table, and I see all the other executives nodding their heads affirmatively. "Yes. Bad decision."

STRATEGIC HINT FOR YOUR CONSIDERATION

IRAC (for issue, rule, application, and conclusion) is a legal decision making technique, which breaks down as follows: What's the issue? What's the rule? Apply the rule. And reach a logical conclusion based on the facts.

Excuse me, chief. "Gordon, you've a thought?" Do we really believe the root cause of this tragedy is a bad decision by the employee? "Well, Gordon, I think he made a bad decision. His boss thinks he made a bad decision. The investigators think he made a bad decision. You think it could be something different?" I think the bad decision is the proximate cause. But the real problem lying in wait might be something different. "What could that be?" Have you ever trained your personnel on how to make good decisions? "Well, of course we have." Show me the class. Show me the curriculum. When you hire people, what's their training agenda? Show me the class on decision making. Organizations rarely offer their people a decision making process. "Well, why should we?" At that point, I remind my clients of the value in performing risk/frequency matrixes. Most of what you do, you're doing right because most of what you do is in the upper and lower right boxes. You've done it before. It's high frequency. You have the memory markers, the behavioral scripts. What happens when you put your people in high-risk, low-frequency activities where they have time to think (critical tasks)? Have you instilled a systematic approach on thinking things through for your people? Do your people have a checklist on how to think things through for these critical tasks? The answer is no unless you're a high-reliability organization. These enterprises deploy sequential and coherent decision making processes for discretionary time activities that are rooted in initial and ongoing training.

You must offer your people a sound decision making process. The California Highway Patrol (CHP) in 1973 didn't provide me a decision making proccss. They taught me how to adeptly perform my duties, but it was technique and activity specific. I spent three years in graduate school where I learned theories and principles of risk management, but I didn't receive a decision making process. Do you know where I received my decision making process? It was in law school. They taught me a technique called IRAC, which breaks down as follows: What's the issue? What's the rule? Apply the rule. And reach a logical conclusion based on the facts. After four years of law school, that's the way my hard drive was programmed to think. "Mr. Graham what's the legal issue here?" "Mr. Graham would you please cite the correct rule of law?" "Mr. Graham would you please apply the rule of law to the facts at hand?" "Mr. Graham would you please reach a logical conclusion based on the facts?"

STRATEGIC HINT FOR YOUR CONSIDERATION

Step one is the most difficult part of the decision making process: What's going on here, and what am I being asked to think about or decide? Listen to what's being communicated to you, and ask clarifying questions.

And for four years, I learned how to think: IRAC, IRAC, IRAC. That was my decision making process. When I opened up the Law Offices of Gordon Graham in 1982, I would mentally perform this process when new clients were speaking: What's the issue? What's the rule? Apply the rule. Reach a logical conclusion. IRAC. It worked for me as a new lawyer. Please don't forget in 1982, I was also a new sergeant. Everything I did since 1973 as a cop I'd done before. I had the memory markers, the behavioral scripts, the mental models. Frequency was on my side. Now I'm a new sergeant, and I'm being exposed to unfamiliar situations and novel experiences. Wait a minute. I wonder if IRAC will work for me as a new sergeant. And it did, but it wasn't perfect. So I kept modifying it, and, in 1985, I devised a decision making process founded on IRAC.

We'll be examining this ten-step decision making process shortly. It's actually eight steps with two follow-up steps, but let's call it a ten-step process. The first step is the most important and warrants additional examination. It involves identifying what needs to be decided. If you can successfully navigate step one, steps two through ten will fall into place. The most difficult part of the decision making process is step one. What's going on here? What am I being asked to think about or decide? If it's preservation of life or property, you skip all the way to step eight, and you preserve life or property. But short of preservation, we start with step one: What's going on here? What am I being asked to do? What's this activity all about? Listen to what's being communicated to you. Ask clarifying questions as necessary. Here's a big one: Don't let your recognition-primed decision making (RPD) get in the way. "Gordon, I thought our RPD was my friend." RPD is your friend, but your RPD can set you up for a nasty phenomenon known as cognitive lock. A great research piece on cognitive lock is, oddly enough, on Wikipedia. For our limited purposes, cognitive lock is where the brain makes up its mind based on partial information. Here's the problem with the brain. Once it starts making up its mind, it hates to change its mind. Cognitive lock can be your coup de grâce.

STRATEGIC HINT FOR YOUR CONSIDERATION

Be aware that cognitive lock can be your coup de grâce because the brain frequently makes up its mind on partial information, and once it formulates a decision, it resists changing directions.

I encourage cops to read a great book on cognitive lock by Erin Torneo et al., called *Picking Cotton*. It's not on my recommended reading list because my list isn't specific to any type of profession. *Picking Cotton* involves a violent sexual assault in North Carolina back in the late 1980s. Street cops show up. They're talking to the victim. Sex detectives show up. The victim is lucid and talking clearly so they start asking her questions. Shortly into their investigation, one sex detective turns to the other detective and says: "I think I know who did this." "What?" "Yes. Same

Modus Operandi. I've dealt with this guy before. He doesn't live that far from here." "Well, there's no harm in talking to him." So they go out to talk to the suspect. He sees the detectives and runs. Does that further prove his guilt? When they catch him, he lies about where he was during the commission of the assault. She identifies the suspect, and they secure a conviction. DNA was analyzed fifteen years later, and it wasn't him. In retrospect, there was exculpatory evidence on scene that was ignored because the detectives had already made up their minds.

Another great book on cognitive lock that's on my recommended reading list is *How Doctors Think by* Jerome Groopman. The first chapter asserts doctors are killing a lot of people but not because they're practicing bad medicine. Patients are dying because doctors are addressing the wrong problems. A typical doctor's visit goes as follows: You have a 9:00 am appointment, and the front desk announces the doctor will see you at 9:45 am. The medical assistant takes you from the waiting area where he weighs you, takes notes, and then asks you to wait for the doctor in a private room. After several minutes, you hear the shuffling papers as well as a knock on the door and then the doctor enters. "Well, Mary." "What brings you in today?" Dr. Groopman lays it out very clearly. The second you start talking, the doctor starts her mental stopwatch. You have eighteen seconds to make your case because your doctor must see sixty patients that day to break even financially. Here's the bad news: You start talking about your sore throat, and guess what? The previous patient had the same symptoms. Therefore, it must be a common cold circulating in the community. If your doctor would've taken the time to ask two more questions, she could've ruled out common cold and properly identified the root problem: rheumatoid arthritis. Once again, cognitive lock.

RPD and cognitive lock can adversely impact your decision making. The same can be said for not evaluating the ethical considerations of your decisions. Webster defines ethics as the discipline of dealing with what's good and bad with moral duty and obligation as well as a set of moral principles or values. Ethical is defined as conforming to accepted professional standards of conduct. Our great country has been sliding down the slippery slope of decreasing ethics and integrity for decades. Not taking ethics seriously has eroded public confidence. We have to rethink how ethical behavior is achieved and providing an annual ethics class (lawyer thinking) isn't the answer. Many organizations use such classes to produce a document certifying their employees have been trained on this behavioral condition. Having such a document doesn't mean much to me. Ethics training must employ the principles of risk management. Number one: If you want to maximize ethical behavior, it starts by hiring people who have integrity. This step requires comprehensive background investigations. Spending wisely on this process is absolutely necessary.

STRATEGIC HINT FOR YOUR CONSIDERATION

Ethics is the discipline of dealing with what's good and bad with moral duty and obligation as well as a set of moral principles or values. Ethical is conforming to accepted professional standards of conduct.

Past habits are indicators of future habits, and the best predictor of future behavior is past behavior (precept one). You can't train the immoral to be moral, so filter them out through background investigations. Number two: After the hiring process, you must train your good new people on ethics and sound decision making. Some next generation employees have substantially different values, and that requires a retuning of their hard drives. Classes on ethics during initial training are important, but it isn't the total answer. Number three: Recognize every activity performed by water/wastewater professionals has ethical considerations.

With this in mind, each training class must have meaningful instruction on the ethical considerations of the particular activity being reviewed. It will cost you nothing to make ethics a part of every training class. That means ethics and ethical decision making doesn't stop with initial training. It must be ongoing and frequent. Finally, when rules aren't being followed, there must be disciplinary action. When supervisors or managers ignore wrongful behavior, they've encouraged future wrongful behavior.

Your supervisors and managers must walk around (SBWA and MBWA) to observe what's actually transpiring in your workplace. You must act if you see something that isn't right. The scope of this book isn't to impart the right thing to do in every situation, for your profession is complex, and the number of permutations of possible activities is innumerable. It's my intent to maximize the level of interest in this concept and to instill a mindfulness that ethics is integral in each decision you make. It's also my aim to stress the value of systems and to give you a structured approach to determine if what you're doing is the right thing to do.

STRATEGIC HINT FOR YOUR CONSIDERATION

Ethics and ethical decision making doesn't stop with initial training. All ongoing training classes must have meaningful instruction on the ethical considerations of the particular activity being reviewed.

Here are four rules on ethics that apply to all line-personnel, supervisors, managers, and executives: (1) Always obey the law and follow the policy (systems). If you have law and policy on your side, you're probably in good shape. (2) If it smells bad, it probably is bad. Even if your planned behavior is consistent with law and policy, it might not be the right thing to do. It's essential that you give it the smell test, both personally and externally. How will it read in the paper tomorrow is a necessary consideration. This rule only applies if you have discretion in what you're doing. In water/wastewater operations, some of your functions are mandatory, meaning you shall perform the activity in a certain way. If you're involved in such an activity, follow the letter of the law or policy regardless of the smell. (3) When questioned after the fact, always be up front and honest. Americans are forgiving but only if you're honest about what happened. Your profession is complex, and mistakes will happen. Don't compound the mistake with a cover up. (4) Ethical actions speak louder than

Graham's Rules for Improved Decision Making (GRIDM©)		
GRIDM© #1	➲	Identify and clarify the issue.
GRIDM© #2	➲	Is there discretionary time or not?
GRIDM© #3	➲	Am I able to address this issue?
GRIDM© #4	➲	What's our current policy regarding the involved issue?
GRIDM© #5	➲	What's our past practice regarding this issue?
GRIDM© #6	➲	Is it the right thing to do under the circumstances?
GRIDM© #7	➲	What are the potential consequences of my decision?
GRIDM© #8	➲	Act!
GRIDM© #9	➲	Document as necessary.
GRIDM© #10	➲	Learn from and share your experiences.

FIGURE 10.1 Graham's rules for improved decision making (GRIDM©). Source: Graham Research Consultants, LLC.

ethical words. You're a leader in your organization and profession, and you must set the proper example. You must act like a professional when no one's looking and when there's no chance your behavior will be noticed. Teach your people how to think, how to do things, and how to know their actions are ethical.

The cornerstone for maximizing employee decision making is a comprehensive and deliberative process comprising rules and processes (systems). I created ten such rules, called GRIDM© (Graham's rules for improved decision making), to help improve the quality of employee decision making so their activities are performed consistently, constantly, and accurately throughout the organization. This type of thinking goes back to the risk/frequency matrix, which ensures your employees are equipped to adeptly handle those activities that are high risk, low frequency, with time to think (critical tasks). GRIDM© is based on the five pillars of success (people, policy, training, supervision, and discipline) and is illustrated in Figure 10.1. Let's examine these rules in detail.

GRIDM© #1

Identify and Clarify the Issue

Step one is the most important step in the process. If there's a preservation of life or property issue, immediately move to step eight and act. Otherwise, ask the following questions: What's going on in this activity? And what am I being asked to do? You can't make the right decision if you're addressing the wrong problem. Listen to what's being communicated to you, and ask clarifying questions. Don't let recognition-prime decision making (RPD) get in the way. RPD can generate cognitive lock. This phenomenon occurs where we make up our mind with little information. That's a problem lying in wait. Study after study has demonstrated the more time you spend identifying what's really going on, the higher the probability you'll make a good decision. Check out www.theinvisiblegorilla.com for some interesting thoughts on this issue, and how it applies to your water/wastewater operations. One last thought: I'm concerned some people will view this request to think as an opportunity to excessively delay performance of given activities. That's not acceptable. You have a job to do so please do your job.

GRIDM© #2

Is There Discretionary Time or Not?

This step is important. If you have discretionary time on a low-frequency activity, then use it to think the issue through using the next five steps of the process. Failure to utilize discretionary time when available is overrepresented in subsequent problems (tragedies). There's no excuse for a poor decision when there's time to think the decision through. Those activities that are non-discretionary time require regular and ongoing training. You're responsible for properly handling these activities should they occur. Most activities, however, give us time to think and allow us to proceed to the next five steps that encapsulate the decision making process. Step two is easy. Do I have time to think? If the answer is yes, proceed to step three. If you don't have time to think, you should've picked that up on your risk/frequency matrix (upper left portion of top-left box) and performed daily training on those activities.

STRATEGIC HINT FOR YOUR CONSIDERATION

Failure to utilize discretionary time when available is overrepresented in subsequent problems (tragedies). There's no excuse for a poor decision when there's time to think the decision through.

GRIDM© #3

Am I Able to Address This Issue?

Step three: Do I have jurisdiction? If the answer is yes, then proceed to steps four, five, six, and seven. If the answer is no, meaning it's a public works or fire department issue, then redirect the issue to someone who has jurisdiction. And whenever possible, follow-through to make sure the issue was properly addressed. Following through is called closing the loop in the realm of customer service. It's an excellent technique for creating loyal customers and generating community goodwill through the WOW factor.

GRIDM© #4

What's Our Current Policy (System) Regarding the Involved Activity, and What Does Our Manual Say about It?

Step one: What's going on? Step two: Do I have time to think? Step three: Do I have jurisdiction? And if the answers for two and three are yes, then proceed to steps four, five, six, and seven. Step four: Get in the habit of looking up policy on low-frequency activities. You don't have to look up policy on high-frequency activities because you're doing them all the time. If you weren't doing them right, you'd already know. Look up policy on low-frequency activities and follow the policy. The easiest way to lose a case is to have a policy in place that stipulates one way to perform the activity, and you performed it a different way. If somebody suffers harm or a loss in that scenario, it's instant liability. If there's no policy specific to what you're trying to do, then refer to your mission statement and comport your behavior around the values and vision of your organization. And for the executives, make sure you have a mission statement: What are you all about as an organization? One last thought on this step: When newer colleagues inquire how to do something, teach them how to look it up.

STRATEGIC HINT FOR YOUR CONSIDERATION

Following through is called closing the loop in the realm of customer service. It's an excellent technique for creating loyal customers and generating community goodwill through the WOW factor.

GRIDM© #5

What's Our Past Practice Regarding This Issue?

Step five: past practice. You may have never experienced this activity before, but someone else in your organization likely has memory markers on its properly handling. Use this discretionary time to ask someone who's done this activity before to ensure your behavior today is consistent with past practice. Failure to treat people consistently is the easiest way to lose community goodwill and customer loyalty. If you're going to deviate from the norm, you must have specific, articulable facts (SAF) to justify this deviation and defend it downstream. Managers and supervisors have a key role in making sure organizational activities are performed consistently by all employees.

GRIDM© #6

Is It the Right Thing to Do Under the Circumstances? What Are the Ethical Considerations?

Every activity has ethical considerations, so every decision must include an ethical analysis. Let's recap. Step one: What's going on? Step two: Do I have time? Step three: Do I have jurisdiction? If the answer to two and three are yes, then step four: What's the policy? Step five: What's our past practice? Step six: Am I doing the right thing here? Not just doing things right, but am I doing the right thing?

Remember, there are four considerations to contemplate when questioning whether your actions are right or ethical: (1) Always follow the law, and always follow the policy. If you have state law and organizational policy on your side, then you're usually headed in the right direction. But don't stop the analysis there. (2) Always give it the smell test, and let me give you a hint: If it smells bad today, it's going to smell worse tomorrow. You can only give it the smell test when you have discretion. When the law says shall or your policy says shall, that's the way it must be done.

STRATEGIC HINT FOR YOUR CONSIDERATION

Failure to treat people consistently eradicates community goodwill and customer loyalty. If you're going to deviate from the norm, you must have specific, articulable facts to justify this deviation and defend it downstream.

STRATEGIC HINT FOR YOUR CONSIDERATION

The easiest way to lose a case is to have a policy in place that stipulates one way to perform the activity, and you performed it a different way. If somebody suffers harm or a loss in that scenario, it's instant liability.

(3) We're going to make mistakes. Your profession involves complex activities. You know what I've learned in life? You can make mistakes and still keep your job so long as you tell the truth. Never compound the mistake with a cover up. I talk to human resources personnel around our great country who say the following: "Gordon, most of the people we're terminating today aren't being terminated for the activity that led to the investigation. They're being terminated for lying during the investigation." Always be up front and honest. Finally, (4) ethical actions speak louder than ethical words. What you do is important and ethical. Practice what you preach, and always be up front and honest. Without the public trust, you have absolutely nothing.

GRIDM© #7

What Are the Potential Consequences of My Decision?

Step seven: the consequence analysis. Smart people consider consequences before making decisions. Misinformed people only consider consequences when consequences arise. By then, it's too late because you're in Lawyerville. Consequences must be part of your decision making process. You must understand and analyze potential consequences in advance of your decision. Consequences include intended, unintended, short-term, and long-term issues. If I do X, how will it end up? If I don't do X, how will it end up? What's the impact on the customer, my coworkers, my organization, and my profession?

You must answer those questions before you act. That's a big issue with newer employees, as they may not understand the gravity of what they're doing. I'm frequently hired to represent employees who are in trouble for not considering the consequences prior to acting. If you're involved in an ethical dilemma, don't make the call yourself. Ask a coworker or supervisor for their advice. If you don't feel comfortable talking about your planned behavior with others, then perhaps you're headed in the wrong direction. Always use your discretionary time to evaluate the consequences of your decisions.

GRIDM© #8

Act! And if the Activity Involves Preservation of Life or Property, Act Quickly

Step eight: act. Make and implement your decision. Do something. If not a preservation of life or property activity, recognize that it's not too late to go back to step one in the process to assure you're still headed in the right direction. It's much easier to start over than to undo something that was done incorrectly.

STRATEGIC HINT FOR YOUR CONSIDERATION

Before performing any discretionary time activity, you must understand and analyze your decision making process through the lens of intended, unintended, short-term, and long-term consequences.

GRIDM© #9

Document as Necessary (the Lawyer in Me)

Step nine is the lawyer in me: The importance of incident documentation, recordkeeping, and detailing why you did, what you did, and when you did it. Recordkeeping and report writing are core components in this process and must be performed as the activity develops. Don't think that you'll remember why you did, what you did, and when you did it three years ago.

GRIDM© #10

Learn from and Share Your Experiences (the Risk Manager in Me)

Step ten is the risk manager in me. If you learn something new, share it with your colleagues. Learn and share. Again, going back to my vexation on institutional knowledge, we aren't sharing our mistakes or sharing our successes. As a result, we keep repeating the same mistakes. If you learn something, share this new memory marker with your colleagues so all can benefit from this knowledge. I'm pretty popular in the fire service and law enforcement. Here's some numbers: There's roughly one-million cops and one-million firefighters in the United States. Who do I represent more in my law office? Cops, and it's not even close. Every time I see a firefighter in my law office, I know I'm going to see fifty cops before I see another firefighter. "Are firefighters smarter than cops?" No, but they learn differently. Let me explain.

What's going on right now in your local police department? You have a cop learning something on a domestic violence call. "Wow. I didn't know that." So she clears the call, and goes where? To the next call, which is an overturned vehicle down an embankment. The tow truck is already on scene and pulling the vehicle to the road. The vehicle is in a rut so the tow truck driver puts a little more pressure on the cable. The cable snaps and rips the mirror off the tow truck. The cop is standing on the other side saying: "Wow. If I would've been standing over there, I could've been hurt." She learned something and where does she go? To her next call. Cops learn individually. Firefighters learn collectively. That's what I want in water/wastewater operations. Collective learning prevents tragedies, whereas individual learning does not. When you learn something, share it with your peers.

FINAL COMMENTS

There are many decision making processes available for your use. If you're a fan of WIN (what's important now), use that process. If you're a fan of John Boyd and his OODA loop (observe, orient, decide, and act), use that one. If you're a military veteran, and they gave you a decision making process, use that one. If you don't have a decision making process, use mine. I like my process for two reasons. (1) It's worked for me throughout my career, and (2) it incorporates ethics, which must be an integral part of any decision. Implanting ethics throughout your organization involves a four-step process. Step one: Hire people who have integrity, which correlates to the people pillar. Background investigations are the best predictor of future behavior because they document past behavior. Step two: After hiring good people, make sure they understand the ethics of your profession. Just because they have integrity, doesn't mean they understand the ethical considerations of why you do what you do in water/wastewater operations. Early on, they need classes on ethics, and these classes must be ongoing and frequent throughout their career. Step three: Decentralize ethics training. Every training class must include an ethical component on the specific activity being taught. Finally, step four: When people don't follow the rules, it must be addressed through discipline. Every time you ignore inappropriate behavior, you've ratified this type of behavior and encouraged future inappropriate behavior. When your people cross the bright line of ethics and integrity, that's a big deal and must be addressed.

That completes our journey toward *real* risk management, but this destination is a base camp for further exploration. We'll accentuate that point in the ensuing conclusion where we'll recapitulate our learning objectives and introduce Admiral Hyman Rickover and his revolutionary thoughts on *real* risk management and high-reliability organizations.

Chapter Takeaway on Media Relations/Crisis Communication

Please refer to the media relations/crisis communication material in the Addendum section and ask yourself how you would address the following scenario:

A neighborhood fire destroyed four homes and a large portion of an apartment complex last night. Local television stations interrupted their scheduled evening shows to report on this four-alarm tragedy. The incident commander's effective leadership and heroism of her firefighters were on full display. As the scene unfolded, one of the captains was interviewed who advised the fire could've been minimized if a nearby fire hydrant was operational. Reporters then began asking neighborhood residents about the fire hydrant. You're stunned with their response. To a person, they all say the fire hydrant had been inoperable since it was damaged by a vehicle nearly three months ago. They then advise your organization refused to fix it even though they complained to your line-personnel whenever they were working in the neighborhood or reading meters.

What? That can't be true. You call your field superintendent while hearing these reports, and she confirmed the hydrant was repaired within ten days of its damage. That makes sense because your organization has a policy to prioritize hydrant repairs. The superintendent went on to advise the policy requires pressure testing by the fire department before reactivation. She also tells you the resident complaints were correct, and she defended her line-personnel for telling the truth. "It's not our problem. We did our part. The rest is up to the fire department." How could our employees think such a response was proper? Yes, it was technically correct but certainly not consistent with our organizational mission of providing meaningful customer service.

Your superintendent also confirmed she promptly submitted a request to the fire department to pressure test the hydrant, but she didn't tell her counterpart the hydrant would be inoperable until his crew performed the test. We have had a great relationship with the fire department, but we never communicated the urgency of performing these pressure tests. Why? It gets worse. She only made one call, and her line-personnel didn't think it necessary to alert her of the constant complaints because they followed policy. How could we be so disconnected to our community? Why didn't we have a decision making process so situations like this one could be properly and consistently addressed?

The news reporters from all the local stations have found their villain, and it's our organization. You're frustrated and annoyed because you work for a stellar organization, and you have good people with good policy. You previously considered a decision making process for your employees to follow, but you rejected the idea as unnecessary. Where's the common sense? Why do we need to train our employees on follow-through? And then you reflect on how it looks to the public. No one closed the loop with the fire department. No one communicated the status with the residents. And no one took the initiative to communicate the numerous complaints to their supervisor.

We now have a community tragedy, inevitable civil liability, and organizational embarrassment. Without knowing it, the respected incident commander articulated both the problem and the blame, and both hit your organization. The media is blaming us and demanding answers. They're already camped out at our headquarters filming a live feed. You're preparing to address the situation in less than an hour with an impromptu press conference, and you know it will be adversarial. How do you frame the messaging, and what do you say to address public anger while also protecting your organization from incriminating statements that will be used against you in the inevitable civil litigation that will follow?

Conclusion
Signing Off from Gordon Graham

A Base Camp for More Journeys

2019 Photo Courtesy of San Diego County Water Authority

STRATEGIC HINT FOR YOUR CONSIDERATION

High-reliability organizations intrinsically realize their tragedies are derived from the human condition, and that actuality necessitates continual improvement of resilient systems.

STRATEGIC HINT FOR YOUR CONSIDERATION

Real risk management embodies a constant state of enterprise vigilance as a means to prevent tragedies, which we define as injury or death to personnel, litigation, embarrassment, investigations, and criminal filings.

We've reached our destination of *real* risk management, and I'd like to thank you for joining me on this journey. It's my sincere hope that you view our destination not as an end point but as a base camp for more exploration in this immeasurable field that spans everything we do in life. For our final section, we'll coalesce the book's core concepts and manifest its key learning objectives. We'll then apply lessons learned through an introduction of Admiral Hyman Rickover and his development of the US Navy nuclear submarine fleet. Rickover is known as the father of the nuclear navy and earned his moniker by transforming the most nascent, complex, and dangerous segment of the US Navy into a high-reliability organization that exemplified *real* risk management. Let's begin with our synopsis.

Real risk management is aptly described as an inexorable pursuit by high-reliability organizations to do the right thing, the right way every time. My philosophy embodies a constant state of enterprise vigilance as a means to prevent tragedies, which we define as injury or death to personnel, civil liability, organizational embarrassment, internal investigations, and criminal filings. High-reliability organizations intrinsically realize their tragedies are derived from the human condition, and that actuality necessitates continual improvement of resilient systems (i.e., control measures, policies, processes, procedures, checklists, rules, and protocols) that are properly designed, kept up to date, and fully implemented. Preventing tragedies through a system approach fosters learning opportunities on both near misses and occasional setbacks because organizational emphasis isn't on assigning employee blame but on system improvement. This approach, however, is ineffective without a culture that drives employee understanding, adherence, questioning, and enforcement of prescribed systems. A successful culture is predicated on employees believing and embracing organizational systems and the underlying rationale behind them. Culture must correlate all behavioral activities to enterprise mission and purpose. Systems and culture are self-reinforcing, with discipline providing the adhesion. When utilized properly, these interdependent components galvanize high-reliability organizations to do the right thing, the right way every time. And that ties back to *real* risk management and the reason these enterprises are underrepresented in tragedies as well as overrepresented in employee fitness and customer loyalty.

STRATEGIC HINT FOR YOUR CONSIDERATION

Your systems will be ineffective unless they integrate the root causational factors of your tragedies. This step requires investigative skill, intellectual curiosity, and fluid communication from all organizational levels.

The mechanics of *real* risk management comprise overlapping and sequential theories derived from three fundamental precepts: (1) Past tragedies are predictive of future tragedies; (2) there's always new ways to prevent tragedies through system improvement; and (3) identifiable risks are manageable risks. One theory yielded from these precepts is RPM (recognition, prioritization, and mobilization). High-reliability organizations utilize RPM by recognizing and prioritizing risks according to probable frequency, severity, and time to think. The recognition component involves risk assessments, which are ongoing analyses of your tragedies and near misses including those of your peers as well as related industry trends. Your risk assessments, through precept one, will confirm most tragedies are caused by employee errors and, therefore, can be proactively managed. A classification system, known as ten families, augments your risk assessments by logically cataloging the thousands of risks facing your organization into understandable and workable categories. This directional support ensures your risk assessments, which must be completed for every job description, are laser-focused on the right risks and not superfluous ones.

The prioritization component is achieved through risk/frequency matrixes intended to draw attention on those risks signifying the greatest impact and likelihood of tragedy. This matrix ranks all job description activities into four grids and places priority on high-risk, low-frequency activities (core critical tasks and critical tasks) that are the source of most organizational tragedies. Like risk assessments, risk/frequency matrixes are an investigative tool assigned to the ten families classification system. The recognition and prioritization components, through precept three, form the data elements for the final component of RPM, which is the mobilization or action phase.

The mobilization (action) component, through precept two, comprises resilient systems that are properly designed, up to date, and fully implemented. The design of resilient systems is built around exhaustive risk assessments and deliberative risk/frequency matrixes. Your systems, however, will be ineffective unless they integrate the root causational factors of your tragedies. This step requires investigative skill, intellectual curiosity, and a free flow of communication from all organizational levels to identify the real causes and not superfluous causes. The five root causes involved in tragedies include lack of quality people, poor or no policy, inadequate or no training, inept supervision, and an absence of organizational discipline.

STRATEGIC HINT FOR YOUR CONSIDERATION

Prudent decision making is an essential component of real risk management. Impulsive and injudicious decisions can have dramatic and permanent consequences that lead to tragedies and loss of community goodwill.

Since tragedies are caused by root causational factors, their prevention requires the adjustment and correction of these factors. High-reliability organizations convert their problem factors into solution factors. The outcome is five pillars of success and its focus on getting and keeping good people, building good policy, assuring great initial and ongoing training, providing adequate supervision, and having a viable discipline process to address arrogance, ignorance, and complacency with fair and impartial rulings. Tragedies are prevented because employees are doing the right thing, the right way every time, and that correlates back to resilient systems and a culture of discipline.

Prudent and ethical decision making is an essential component of *real* risk management. Impulsive and injudicious decisions can have dramatic and permanent consequences that lead to tragedies and loss of community goodwill. High-reliability organizations build resilient systems and impart deliberative and ethical decision making processes to assist employees in improving their operational performance. An assiduous decision making structure encompasses a methodical checklist of five separate and distinct issues (themes) that, when put together, allow employees to analyze all of their activities using risk management, systems, customer service, accountability, and integrity. Each activity is analyzed (if there's discretionary time) with these questions: What's the risk involved in this activity, and how can I best manage that risk? What's our system (policies, procedures, processes, protocols, checklists, and rules), and how can I best assure its implementation? Is there a customer service component here, and, if so, how can I maximize customer service on this activity? Who's accountable for what on this activity? What are the issues of integrity involved in this specific activity? Answering these questions will ensure consistent outcomes in quality, responsiveness, and conformance. This process is based on discipline, which boosts accountability and cures mediocrity. Discipline safeguards ethical and sound decision making by ensuring employees think through the consequences of their actions.

Real risk management represents various organizational layers of defense, reminders, and support to prevent tragedies by enabling employees to do the right thing, the right way every time. It takes a good person to be a good professional, but being a good person isn't enough. In order to be thoroughly ethical and proficient, water/wastewater professionals must have a good, workable policy manual that allows good people to know how to act in a given situation. Additionally, good people must be fully trained to perform every aspect of their rightful work.

STRATEGIC HINT FOR YOUR CONSIDERATION

A management approach of system reliance fosters learning opportunities on both near misses and occasional setbacks because organizational emphasis isn't on assigning employee blame but on system improvement.

Training is an everyday event, and it must focus on the high-risk, low-frequency activities, particularly those with no time to think. Supervisors and managers must be accountable to ensure employees are doing the right thing, the right way every time. When they're not, that needs to be addressed through discipline, not because of consequences but because of deviation from policy. Performing assigned activities the right way, the first time, while treating all with dignity and respect, should be the goal of every employee. Any failure from one or more of the five pillars will eventually enable your problems lying in wait to align, propel, and exit through the holes in the Swiss cheese. At that point tragedy occurs, lawyers take over, and your problems lying in wait are identified after the fact.

I'd like to now transition and crystallize all these learning objectives by introducing Admiral Hyman Rickover (1900–1986) and his legendary rules and management philosophy. I was fortunate to be introduced to Rickover's work at the Institute of Safety & Systems Management (ISSM) and was overwhelmed by the contributions this valiant immigrant afforded our great country.

Admiral Rickover directed the building of our nuclear submarine fleet that protected the free world and simultaneously achieved an outstanding safety record. He was one of the most successful and controversial public managers and military leaders of the twentieth century. His accomplishments are legendary. In three years, Rickover's team designed and built the first nuclear submarine, the USS *Nautilus*, an amazing feat of engineering given it involved the development of the first controlled nuclear reactor in a naval boat. The *Nautilus* not only transformed submarine warfare; it also laid the groundwork for our fleet of nuclear aircraft carriers and cruisers (which were also built by Rickover and his team).

Rickover's extraordinary safety and reliability record were the results of innovation, diligence, and continual improvement. He developed specific rules to achieve success (read: safe operations and deployment ready) known colloquially as the Seven Rules of Rickover. One of the goals at ISSM was to learn how his rules could be made applicable to other branches of the US military. As I sat there forty-plus years ago, I wondered if these rules applied to my complex, high-risk job at the California Highway Patrol (CHP). Rickover's rules apply to all high-reliability organizations, and when taken seriously will improve the quality of your operations. Let's take a look at each of these rules and examine how they apply to the complex world of water/wastewater operations as well as how they tie back to my rules on *real* risk management.

STRATEGIC HINT FOR YOUR CONSIDERATION

Continuous improvement must define the way you do business. Anything you can quantify, and anything you can measure, has to be identified, and you must be constantly searching for the next best way.

RULE ONE

You Must Have a Rising Standard of Quality over Time, Well beyond What's Required by Any Minimum Standard

You have to continually improve at what you do. Minimum standards are just that: minimum standards. Your profession deserves better than minimum standards. The communities you serve deserve better than minimum standards. Your personnel deserve better than minimum standards. You must constantly be looking for a better way to do things. The status quo days are no longer acceptable. Sadly, I see a lot of status quo around our great country. Continuous improvement must be part of the way you do business. Anything you can quantify, and anything you can measure, has to be identified, and you must constantly be searching for the next best way. When you find the next best way, you must commence the search for the next best way. I'm not talking about change for change's sake, but a bona fide effort to continually improve your organization.

RULE TWO

People Running Complex Systems Should Be Highly Capable

STRATEGIC HINT FOR YOUR CONSIDERATION

Organizational success runs through your supervisors. They must have the fortitude to drive behavior that conforms to your systems and the perspicuity to explain the rationale behind these systems.

Successful operations require people who know how to think. Fifty years ago, you didn't have to graduate with distinction to be a cop, firefighter, or water/wastewater operator. Back then, you had to be competent and a hard worker. While these attributes are still important, we must recognize the vast changes to your profession. Technology, equipment, strategies, and tactics involved in providing water/wastewater to your community have become more complex. If you hire people who can't think things through, you're en route to disaster. Such individuals may end up being the cause of future tragedy, and your organization, nor your community, can afford that bargain. We've learned this lesson time and time again, but somehow we seem to forget it all too often. Please don't tell me that you have nothing to do with the hiring process. Each of you has a role in recruitment, and each of you has a role in the probationary process of new employees. You have to take your role in these processes seriously. I could tell you stories about water-related entities from around the United States that failed to filter out a bad actor and paid the price. Every dollar you spend filtering out bad actors has the potential to save you millions of dollars in downstream litigation. You must invest the necessary time and resources to onboard highly capable people.

RULE THREE

Supervisors Have to Face Bad News When It Comes, and Take Problems to a Level High Enough to Fix Those Problems

When you take an honest look at your organizational tragedies (i.e., lawsuits, injuries, deaths, embarrassments, internal investigations, and criminal filings), so many are attributed to supervisors not behaving like supervisors. The primary mission of a supervisor is systems enforcement. If you promote people who either can't or won't enforce policy, you're en route to tragedy. To be sure, the transition from line employee to supervisor is a difficult one, but the people chosen to be supervisors must understand the importance of their job. We have too many people who call themselves supervisors who've never made a successful transition from buddy to boss. Show me a tragedy in any operation, and I will show you the fingerprints of a supervisor not behaving like a supervisor; or a supervisor who tried to do her job and wasn't supported by management. Organizational success runs through your supervisors. They must have the fortitude to drive behavior that conforms to your systems and the perspicuity to explain the rationale behind these systems. Equally important, it requires speaking truth to power and applying measured discipline when rules are violated.

STRATEGIC HINT FOR YOUR CONSIDERATION

The profession you've chosen is filled with risk, and there's always a potential for the unthinkable to occur in your workplace. Solid, realistic, ongoing, and verifiable training (SROVT) is critical for preventing tragedies.

RULE FOUR

You Must Have a Healthy Respect for the Dangers and Risks of Your Particular Job

All of your jobs in water/wastewater operations are high risk in nature, and the consequences for not properly performing activities (tasks, incidents, and events) can be dramatic. Remember the theory of RPM (recognition, prioritization, and mobilization). You must recognize the risks you face and prioritize these risks in terms of potential frequency, severity, and time to think. Then you must mobilize (act) to prevent the identified problems from occurring. The profession you've chosen is filled with risk, and there's always a potential for the unthinkable to occur in your workplace. Solid, realistic, ongoing, and verifiable training (SROVT) is critical for preventing tragedies as well as maximizing employee fitness and customer loyalty.

RULE FIVE

Training Must Be Constant and Rigorous

Every day must be a training day. You must focus the training on activities in every job description that have the highest probability of causing us grief (tragedies). These are the high-risk, low-frequency activities where there's no time to think (core critical tasks). Every job description in your organization has these core critical tasks, and they must be addressed. You have a responsibility to ensure all personnel are fully and adequately trained on the activities that give them no time to think. The same responsibility applies to their understanding the value of thinking things through when time allows. The vast majority of activities your people perform, they're doing right. But when things don't go right, there's always a reason. Where do you need to focus your efforts when you get back to work? James Reason, one of the great writers on risk management, categorizes the different types of errors into knowledge based, rules based, and skills based. He goes on to talk about how errors, lapses, omissions, and mistakes happen in any given workplace as well as the importance of reviewing past errors and instilling control measures (training) to prevent their reoccurrence. Your role is making sure employees in every job description are adequately trained for non-discretionary time activities (core critical tasks), and that they understand the value of thinking things through when they're involved in discretionary time activities. It all correlates to the five pillars of success: people, policy, training, supervision, and discipline.

RULE SIX

Audits and Inspections of All Aspects of Your Operations Are Essential

STRATEGIC HINT FOR YOUR CONSIDERATION

Analysis of past data is the foundation of real risk management. All of you in water/wastewater operations are repeating the same mistakes. You can prevent future mistakes by studying the mistakes of the past.

STRATEGIC HINT FOR YOUR CONSIDERATION

If you take the time to study the life of Admiral Rickover, you'll quickly learn he was widely despised in the Navy because of his insistence on using the audit process as a tool to hold people accountable.

Audits and inspections are an important part of your job in water/wastewater operations. You can't assume everything is going well. You must have control measures (systems) to ensure things are being done the right way every time. This process isn't micro-management. It's called doing your job. You need a feedback loop in every organization. And while I'm ignorant on the internal workings of your specific organization, I've looked at hundreds of enterprises in various professions, and audits are either non-existent or deficient. I call these the ersatz audits where priority is a document certifying all's well, when in reality that's not true. The lack of a robust audit process (formal and informal) is a problem lying in wait and an invitation to tragedy. You're then in the domain of lawyers, and it's too late for action. All you can do is address the consequences. If you take the time to study the life of Admiral Rickover, you'll quickly learn he was widely despised in the Navy because of his insistence on using the audit process as a tool to hold people accountable.

RULE SEVEN

The Organization and Members Thereof Must Have the Ability and Willingness to Learn from Mistakes of the Past

Analysis of past data is the foundation of *real* risk management. All of you in water/wastewater operations are repeating the same mistakes. As I review your profession's lawsuits, injuries and deaths, organizational embarrassments, internal investigations, and even the rare criminal filings, I know you can prevent future mistakes by studying the mistakes you've made in the past. It all condenses down to risk management. There's a tremendous amount of risk in water/wastewater operations. Risk can be eliminated, avoided, shared, controlled, or transferred. For those of you who are executives (today or in the future), you must be evaluating your risks. Each of the preceding techniques is a form of risk management. Every identifiable risk is a manageable risk. Unfortunately, many water/wastewater professionals don't fully comprehend the purpose and value of *real* risk management. Allow me to briefly explain. What are you doing that you shouldn't be doing? What high-risk activities can you transfer to someone else? How can you better control the risk involved in this operation? It all gets down to managing risk. Let's start off with what risk management isn't: *Real* risk management isn't another assigned duty of the finance director, a quick mention during supervisor training, or a safety stuff. Everything we do involves a level of risk, and *real* risk management must be fully inculcated in your organization to ensure you do the right thing, the right way every time.

RICKOVER'S FINAL THOUGHTS

The best way to encapsulate the management philosophy of Admiral Rickover is from his own words. I can't think a more fitting way to conclude our time together than to share one of his last public speeches. You'll soon note his masculine pronoun usage. This usage was reflective of his time but inaccurately reflects today's workplace of highly skilled women leaders. The words that follow are credited to www.govleaders.com, and I encourage you to share this speech with your people. It personifies a high-reliability organization and underscores the brilliance of a visionary military leader who I tribute as an original user of real risk management.

> The [following] text is an excerpt from a speech Rickover delivered at Columbia University in 1982, in which he succinctly outlined his management philosophy. His determination, clarity of purpose, emphasis on developing his people, high standards, and willingness to give his people ownership of their work had to have been very inspiring. He had exceptionally high standards and was known to take some of these same strengths to extremes, however, which no doubt led to his reputation in some circles as being difficult to work for. On that cautionary note, we are pleased to present Rickover's own description of his management style:
>
> Human experience shows that people, not organizations or management systems, get things done. For this reason, subordinates must be given authority and responsibility early in their careers. In this way, they develop quickly and can help the manager

do his work. The manager, of course, remains ultimately responsible and must accept the blame if subordinates make mistakes. As subordinates develop, work should be constantly added so that no one can finish his job. This serves as a prod and a challenge. It brings out their capabilities and frees the manager to assume added responsibilities. As members of the organization become capable of assuming new and more difficult duties, they develop pride in doing the job well. This attitude soon permeates the entire organization. One must permit his people the freedom to seek added work and greater responsibility. In my organization, there are no formal job descriptions or organizational charts. Responsibilities are defined in a general way, so that people are not circumscribed. All are permitted to do as they think best and to go to anyone and anywhere for help. Each person then is limited only by his own ability.

Complex jobs cannot be accomplished effectively with transients. Therefore, a manager must make the work challenging and rewarding so that his people will remain with the organization for many years. This allows it to benefit fully from their knowledge, experience, and corporate memory. The Defense Department does not recognize the need for continuity in important jobs. It rotates officers every few years both at headquarters and in the field. The same applies to their civilian superiors. This system virtually ensures inexperience and non-accountability. By the time an officer has begun to learn a job, it is time for him to rotate. Under this system, incumbents can blame their problems on predecessors. They are assigned to another job before the results of their work become evident. Subordinates cannot be expected to remain committed to a job and perform effectively when they are continuously adapting to a new job or to a new boss.

When doing a job—any job—one must feel that he owns it, and act as though he will remain in the job forever. He must look after his work just as conscientiously, as though it were his own business and his own money. If he feels he is only a temporary custodian, or that the job is just a stepping stone to a higher position, his actions will not take into account the long term interests of the organization. His lack of commitment to the present job will be perceived by those who work for him, and they, likewise, will tend not to care. Too many spend their entire working lives looking for their next job. When one feels he owns his present job and acts that way, he need have no concern about his next job. In accepting responsibility for a job, a person must get directly involved. Every manager has a personal responsibility not only to find problems but to correct them. This responsibility comes before all other obligations, before personal ambition or comfort.

A major flaw in our system of government, and even in industry, is the latitude allowed to do less than is necessary. Too often officials are willing to accept and adapt to situations they know to be wrong. The tendency is to downplay problems instead of actively trying to correct them. Recognizing this, many subordinates give up, contain their views within themselves, and wait for others to take action. When this happens, the manager is deprived of the experience and ideas of subordinates who generally are more knowledgeable than he in their particular areas. A manager must instill in his people an attitude of personal responsibility for seeing a job properly accomplished. Unfortunately, this seems to be declining, particularly in large organizations where responsibility is broadly distributed. To complaints of a job poorly done, one often hears the excuse, "I am not responsible." I believe that is literally correct. The man who takes such a stand in fact is not responsible; he is irresponsible. While he may not be legally liable, or the work may not have been specifically assigned to him, no one involved in a job can divest himself of responsibility for its successful completion.

Unless the individual truly responsible can be identified when something goes wrong, no one has really been responsible. With the advent of modern management theories, it is becoming common for organizations to deal with problems in a collective manner, by dividing programs into subprograms, with no one left responsible for the entire effort. There is also the tendency to establish more and more levels of management, on the theory that this gives better control. These are but different forms of shared responsibility, which easily lead to no one being responsible—a problem that often inheres in large corporations as well as in the Defense Department. When I came to Washington before World War II to head the electrical section of the Bureau of Ships, I found that one man was in charge of design, another of production, a third handled maintenance, while a fourth dealt with fiscal matters. The entire bureau operated that way. It didn't make sense to me. Design problems showed up in production, production errors showed up in maintenance, and financial matters reached into all areas. I changed the system. I made one man responsible for his entire area of equipment—for design, production, maintenance, and contracting. If anything went wrong, I knew exactly at whom to point. I run my present organization on the same principle.

A good manager must have unshakeable determination and tenacity. Deciding what needs to be done is easy, getting it done is more difficult. Good ideas are not adopted automatically. They must be driven into practice with courageous impatience. Once implemented they can be easily overturned or subverted through apathy or lack of follow-up, so a continuous effort is required. Too often, important problems are recognized but no one is willing to sustain the effort needed to solve them. Nothing worthwhile can be accomplished without determination. In the early days of nuclear power, for example, getting approval to build the first nuclear submarine—the Nautilus—was almost as difficult as designing and building it. Many in the Navy opposed building a nuclear submarine. In the same way, the Navy once viewed nuclear-powered aircraft carriers and cruisers as too expensive, despite their obvious advantages of unlimited cruising range and ability to remain at sea without vulnerable support ships. Yet today our nuclear submarine fleet is widely recognized as our nation's most effective deterrent to nuclear war. Our nuclear-powered aircraft carriers and cruisers have proven their worth by defending our interests all over the world—even in remote trouble spots such as the Indian Ocean, where the capability of oil-fired ships would be severely limited by their dependence on fuel supplies.

The man in charge must concern himself with details. If he does not consider them important, neither will his subordinates. Yet "the devil is in the details." It is hard and monotonous to pay attention to seemingly minor matters. In my work, I probably spend about ninety-nine percent of my time on what others may call petty details. Most managers would rather focus on lofty policy matters. But when the details are ignored, the project fails. No infusion of policy or lofty ideals can then correct the situation. To maintain proper control one must have simple and direct means to find out what is going on. There are many ways of doing this; all involve constant drudgery. For this reason those in charge often create "management information systems" designed to extract from the operation the details a busy executive needs to know. Often the process is carried too far. The top official then loses touch with his people and with the work that is actually going on.

Attention to detail does not require a manager to do everything himself. No one can work more than twenty-four hours each day. Therefore to multiply his efforts, he must create an environment where his subordinates can work to their maximum ability. Some management experts advocate strict limits to the number of people reporting to a common superior—generally five to seven. But if one has capable people who require but a few moments of his time during the day, there is no reason to set such arbitrary constraints. Some forty key people report frequently and directly to me. This enables me to keep up with what is going on and makes it possible for them to get fast action. The latter aspect is particularly important. Capable people will not work for long where they cannot get prompt decisions and actions from their superior.

I require frequent reports, both oral and written, from many key people in the nuclear program. These include the commanding officers of our nuclear ships, those in charge of our schools and laboratories, and representatives at manufacturers' plants and commercial shipyards. I insist they report the problems they have found directly to me—and in plain English. This provides them unlimited flexibility in subject matter—something that often is not accommodated in highly structured management systems—and a way to communicate their problems and recommendations to me without having them filtered through others. The Defense Department, with its excessive layers of management, suffers because those at the top who make decisions are generally isolated from their subordinates, who have the first-hand knowledge.

To do a job effectively, one must set priorities. Too many people let their "in" basket set the priorities. On any given day, unimportant but interesting trivia pass through an office; one must not permit these to monopolize his time. The human tendency is to while away time with unimportant matters that do not require mental effort or energy. Since they can be easily resolved, they give a false sense of accomplishment. The manager must exert self-discipline to ensure that his energy is focused where it is truly needed. All work should be checked through an independent and impartial review. In engineering and manufacturing, industry spends large sums on quality control. But the concept of impartial reviews and oversight is important in other areas also. Even the most dedicated individual makes mistakes—and many workers are less than dedicated. I have seen much poor work and sheer nonsense generated in government and in industry because it was not checked properly.

One must create the ability in his staff to generate clear, forceful arguments for opposing viewpoints as well as for their own. Open discussions and disagreements must be encouraged, so that all sides of an issue will be fully explored. Further, important issues should be presented in writing. Nothing so sharpens the thought process as writing down one's arguments. Weaknesses overlooked in oral discussion become painfully obvious on the written page. When important decisions are not documented, one becomes dependent on individual memory, which is quickly lost as people leave or move to other jobs. In my work, it is important to be able to go back a number of years to determine the facts that were considered in arriving at a decision. This makes it easier to resolve new problems by putting them into proper perspective. It also minimizes the risk of repeating past mistakes. Moreover if important communications and actions are not documented clearly, one can never be sure they were understood or even executed.

It is a human inclination to hope things will work out, despite evidence or doubt to the contrary. A successful manager must resist this temptation. This is particularly hard if one has invested much time and energy on a project and thus has come to feel possessive about it. Although it is not easy to admit what a person once thought correct now appears to be wrong, one must discipline himself to face the facts objectively and make the necessary changes—regardless of the consequences to himself. The man in charge must personally set the example in this respect. He must be able, in effect, to "kill his own child" if necessary and must require his subordinates to do likewise. I have had to go to Congress and, because of technical problems, recommended terminating a project that had been funded largely on my say-so. It is not a pleasant task, but one must be brutally objective in his work.

No management system can substitute for hard work. A manager who does not work hard or devote extra effort cannot expect his people to do so. He must set the example. The manager may not be the smartest or the most knowledgeable person, but if he dedicates himself to the job and devotes the required effort, his people will follow his lead. The ideas I have mentioned are not new—previous generations recognized the value of hard work, attention to detail, personal responsibility, and determination. And these, rather than the highly-touted modern management techniques, are still the most important in doing a job. Together they embody a common-sense approach to management, one that cannot be taught by professors of management in a classroom. I am not against business education. A knowledge of accounting, finance, business law, and the like can be of value in a business environment. What I do believe is harmful is the impression often created by those who teach management that one will be able to manage any job by applying certain management techniques together with some simple academic rules of how to manage people and situations.

Addendum

The Ten Families of Risk

Family One	External Risks Risks arising from outside the organization that impact your operations and are the most difficult family of risk to address.
Family Two	Legal and Regulatory Risks Risks arising from the complexity of, or non-compliance with, the legal framework imposed on water-related entities.
Family Three	Strategic Risks Risks arising from the lack of priority setting and business planning leading to a reactive organization that isn't prepared or flexible enough to address unforeseen events.
Family Four	Organizational Risks Risks arising from not clearly defining roles and responsibilities, not demonstrating the value of your organization, and not having monitoring processes in place. You must be concerned with people, policy, training, supervision, and discipline. These factors are often the root cause of tragedies.
Family Five	Operational Risks Risks arising from performing tactical activities (i.e., tasks, incidents, and events) with systems that aren't continuously improved as well as not being properly designed, up to date, and fully implemented.
Family Six	Information Risks Risks arising from untimely, inaccurate, or unreliable information that supports the discharge of roles and responsibilities.
Family Seven	Technology Risks Risks arising from outdated or unreliable information systems where user requirements aren't met and the resulting impact is compromised security measures (internal and external). It's the fastest growing family of risks.
Family Eight	Human Resources Risks Risks arising from work environments that don't receive adequate resource allocations, lack internal mutual trust, forego acceptable performance levels, or suffer from poor transparency and good management. It's the most expensive but easiest family to address.
Family Nine	Financial and Reputational Risks Risks arising from improper budgeting, forecasting, and expenditure controls, including contracting, asset management, internal audits, improper salaries, theft of cash, misappropriation of resources, misuse of overtime, and poor revenue oversight.
Family Ten	Political Risks Risks arising from your dealings with elected personnel. This family should be avoided when possible.
Questions	What are the three greatest risks faced by your organization in each of these families? What are your control measures (systems) to address each of your identified risks?

Source: Graham Research Consultants, LLC.

Five Concurrent Themes for Success	
A Checklist for Doing the Right Thing, the Right Way Every Time	
Theme One	Risk What's the risk involved in this activity, and how can I best manage that risk?
Theme Two	Policy What's our organizational policy (system), and how can I best assure its implementation?
Theme Three	Service Is there a customer service component here, and, if so, how can I maximize customer service on this activity?
Theme Four	Accountability Who's accountable for what on this specific activity?
Theme Five	Integrity What are the issues of integrity involved in this specific activity?

Source: Graham Research Consultants, LLC.

Graham's Rules for Improving Background Investigations (GRIBI©)	
GRIBI© #1	Understand the purpose and importance of good background investigations. The purpose of a background investigation is to find relevant past behaviors and document these behaviors so the hiring decision maker has sufficient information to make an informed decision.
GRIBI© #2	Establish a policy regarding background investigations. Your policy needs to be specific to your organization, and because of the high-risk nature of the activity, it must be reviewed by competent counsel on a regular basis and include waivers specific to the job description you're testing for and specific to the applicant.
GRIBI© #3	Select people who have the knowledge, skill, and desire to conduct background investigations. Not everyone in your organization is qualified to be a background investigator. The job is tedious, requires a lot of documentation and meticulous review, and is intensive in areas that aren't all that exciting. However, the process is critical to the survival of your organization.
GRIBI© #4	The process shouldn't be a secret. The overall background investigation process should be available for inspection by the public and by applicants seeking employment. Showing potential applicants how the process works may discourage those who know their past behaviors will be discovered from applying.
GRIBI© #5	Take the time to do the job right. Background investigations are a discretionary time activity. Your investigators should have the necessary time to gather all the facts about an applicant's past behaviors and present situation.
GRIBI© #6	Utilize all available resources, and don't worry about the cost. Public-sector and private-sector databases should be fully accessed within the applicable laws of your jurisdiction. This process should receive the proper resources to ensure a proper outcome.

GRIBI© #7 Remember the importance of accuracy.
Your investigative package is subject to discovery, and it can't be changed from what you've originally submitted. Full and accurate documentation of each step of the process is essential.

GRIBI© #8 Always proofread your documentation.
Proofreading will reinforce the importance of accuracy and help assure the needs of number seven above. Spend the time to critically review what you've documented prior to submitting it.

GRIBI© #9 Supervisory review is essential.
The background investigation process is high risk, and there are substantial consequences for not doing the activity correctly. Minimize risk and exposure by incorporating knowledgeable supervisors into the review process, not just at the terminus of the investigation, but throughout the process.

GRIBI© #10 Learn from your experiences.
As you learn, share your knowledge with your colleagues. If your newfound knowledge isn't in the organizational policy, make the appropriate recommendations to reflect this better way to perform background investigations.

Source: Graham Research Consultants, LLC.

Waiver for Lateral Hires<?>

I *(Insert Full Name of Applicant)* am making application to become a *(Insert Desired Job Title)* for the *(Insert Your Organization)*. I am currently employed as a *(Insert Current Job Title)* for the *(Insert Current Organization)*. I fully understand that the *(Insert Your Organization)* will perform a complete and thorough background investigation to ensure I have the necessary skills, abilities, and integrity to perform as a *(Insert Desired Job Title)* of and for *(Insert Your Organization)*.

I recognize and understand that this background investigation will include but not be limited to personal history, usage of illegal drugs, criminal misconduct, domestic violence, immoral behavior, and any other behaviors deemed by *(Insert Your Organization)* to be essential for service as a *(Insert Desired Job Title)*. I also fully understand that information learned by the *(Insert Your Organization)* may result in my not being hired. Recognizing all of the above, I hereby give the *(Insert Your Organization)* full and complete permission to disclose the findings and results of this comprehensive background investigation to my current employer, *(Insert Current Organization)*.

I understand that this disclosure may result in adverse consequences for me in my current job, including but not limited to termination from employment, negative reference for future employment, and possible criminal prosecution. I agree to hold the *(Insert Your Organization)* harmless from any and all claims made by me as a result of this release of information. I have initialed each of the above paragraphs and have signed this waiver at the bottom of this page. I fully understand this waiver and have been offered the opportunity to withdraw my application for employment to the *(Insert Your Organization)*.

____

Current Employer/Address of Current Employer

____

Current Human Resources Head/Phone Number

Signed this ____ of ________________, _______ at _______________________, State. ____

____

Applicant Witness

Source: Graham Research Consultants, LLC.

Graham's Rules for Improving Performance Evaluations (GRIPE©)	
GRIPE© #1	Understand the purpose of employee evaluations. Performance evaluations can improve the morale of employees who meet performance expectations and provide fair warning to employees whose performance is unsatisfactory and where improvement is needed to meet standards.
GRIPE© #2	Implement a policy (system) regarding evaluations. Your performance evaluation policy must be specific to each job description within the organization. Well-prepared job descriptions as well as performance goals and expectations are an essential precursor to the evaluation process.
GRIPE© #3	Select people to be supervisors who are capable of being honest with employees. When you're promoting people in your organization, make sure they have the ability to be honest with employees, including during the rating process. Implement a training program for all supervisors and managers so that everyone understands what evaluations are about, the consequences of doing them wrong, and most importantly, how to do them properly.
GRIPE© #4	The process shouldn't be a secret. Make sure all employees are aware of how the process works, including how data is collected and analyzed for use in the evaluation.
GRIPE© #5	Before you put pencil to paper (or fingers on a keyboard), think. Performance evaluations are a high-risk activity, but they're discretionary time. That means you have time to think before you perform the activity. Use this time to review the policy (system) of your organization, and make sure you understand what you're doing.
GRIPE© #6	Be specific during the evaluation process. It's important that all rating be based on specific, articulable facts (SAF) so that the rating is supported and defensible.
GRIPE© #7	Remember the importance of accuracy. Accuracy is essential in a well-prepared performance evaluation, particularly if the evaluation has negative elements that may impact future employment or promotion.
GRIPE© #8	Always proofread your documentation. Minor errors can cause major problems so double-check your evaluation prior to forwarding it to your manager.
GRIPE© #9	If you're right, don't change it. This step is a personal call, but if you change a document to reflect performance traits you know didn't exist, you may be setting yourself up for bigger problems downstream.
GRIPE© #10	Learn from and share your experiences. Share your knowledge with your peers. If your newfound knowledge isn't in the organizational policy, make the appropriate recommendations to reflect this better way to do evaluations.

Source: Graham Research Consultants, LLC.

Using the Annual Performance Evaluation (APE) to Reduce Harassment in Your Workplace

The annual performance evaluation process is much more than giving an employee a piece of paper rating their performance over the last reporting period. This process, if properly used, gives the employer the opportunity to verify the employee's level of knowledge regarding core critical tasks and critical tasks for their position. In particular, verification of the employee's knowledge regarding high-risk, low-frequency activities is crucial. One such area is job-based harassment, which can also lead to future allegations of retaliation. The performance evaluation process can be used not only to check the employee's level of knowledge regarding your organization's policy, but also to identify issues that need to be discussed and mitigated. The following is a template you can incorporate into your performance evaluations. First, the employee and supervisor/HR representative review the harassment policy together. Then, the following series of questions are asked, with the employee's responses documented.

Note: This process should be followed with each employee being evaluated.

Pat, we have just read our (Insert Your Organization) policy regarding job-based harassment in the workplace. I am now required to ask you some questions, and it will be necessary for you to document your response to these questions:

- Do you understand this policy? Yes/No ________ Initial
- Do you have any questions about this policy? Yes/No ________ Initial
- Do you know how to file a complaint should you ever have a problem with harassment, or if you see inappropriate behaviors at work? Yes/No ________ Initial
- If you ever have a problem or concern regarding harassment in the workplace, you can address your concerns with the following resources, both inside and outside our organization. List the resource contacts. Do not use names here, titles only.
- Are you aware of any behaviors going on either in our workplace or outside the workplace that may impact the workplace and are inconsistent with this policy? Yes/No ________ Initial

If the answer to this last query is Yes, then gather the facts of their concern, notify human resources, and commence an investigation pursuant to your harassment policy and procedure. In the closing comments on the evaluation, incorporate the above questions and responses.

For example:

Pat, another excellent year. You built four-million widgets; you do this, this, this, and this well. You need to improve here, here, and here. You manage your time well, you follow safety rules, etc. Additionally, we had a conversation regarding our organizational commitment to the elimination of harassment in the workplace. We read our organization's policy together, and you stated you understood the policy and had no questions. Additionally, you stated you were unaware of any behaviors inconsistent with this policy. Should you ever become aware of behaviors inconsistent with our policy, please do not hesitate to contact me, any other supervisor or manager, or human resources immediately and inform us of the inconsistent behavior.

This approach is an excellent way to eliminate harassment for two reasons: (1) It makes every employee aware that someone is watching out for problems, and (2) Should the annual performance questioning reveal a potential harassment issue, appropriate staff can promptly address it in accordance with your policy. Of course, all of that assumes that you have a good, up-to-date harassment policy, and that you are committed to eliminating this issue from your workplace. It's the right thing to do.

Note: Prior to changing any policy or procedure you now have, please consult your competent legal counsel.

Source: Graham Research Consultants, LLC.

Graham's Rules for Improving Supervisory Performance (GRISP©)*

GRISP© #1	How do you select your supervisory team? Time on-the-job requirement of ten thousand hours. Certification requirements prior to promotion. Testing considerations. Supervisory responsibilities prior to promotion.
GRISP© #2	Some initial training considerations. Training prior to operations, including buddy-to-boss considerations as well as daylight-versus-artificial light decision making (non-discretionary time versus discretionary time). You can't enforce policy (system) if you don't know policy. Bring back the best-of-the-best to help develop your supervisors.

GRISP© #3	Some ongoing training considerations. Some thoughts on continuous learning: Mandatory reading programs are invaluable. The importance of learning from events, both good and bad: You must learn from mistakes prior to tragedy. NPCCR (non-punitive close call reporting) is an excellent tool to deploy.
GRISP© #4	Clearly define your expectations of performance. The primary mission of a supervisor is systems enforcement. Someone has to enforce your organizational policies and procedures.
GRISP© #5	Share with your supervisory team the problems caused by a lack of systems enforcement. When rules aren't followed, bad things can happen. These include death or injury to personnel, civil liability, embarrassment, internal investigations, and even the rare criminal filing.
GRISP© #6	Hey Boss! Back up your supervisors when they make the tough calls. Always remember perception is reality. If supervisors think they won't get the full support of the management after a tough call is made, they'll avoid making the tough calls.
GRISP© #7	You must have an organizational feedback loop to ensure what you say you're doing is in fact being done. Formal audits are essential. Informal audits like managing by walking around (MBWA) and supervising by walking around (SBWA) are good tools to use. The importance of analyzing events that ended up right: Just because activities end up okay doesn't mean you did things right.
GRISP© #8	Give your people credit for what they've done. Prepare them for further promotion. Reinforce the value of their contribution. Praise them in public. Criticize in private. Set the proper example.
GRISP© #9	Constantly look for the next best way. There's always a better way, and we must search for it by sharing memory markers.
GRISP© #10	A thought from Ross Swope. The major cause in the lack of integrity in American Police Officers is mediocrity. This quote applies to all professions, including water/wastewater operations.

Source: Graham Research Consultants, LLC.

Graham's Rules for Appropriate Employee Discipline (GRAED©)

GRAED© #1	Understand the importance of organizational discipline. If you want things to go right, you must be aware of and pay attention to the five pillars of success. Take away one of these pillars, and bad things will happen. The fifth pillar of success, discipline, is often not taken seriously, resulting in major consequences (tragedies).
GRAED© #2	Develop good organizational policies and procedures (systems). Well-written policies give guidance to employees on how to achieve the organizational expectation of doing the right thing, the right way every time. Without a written policy, you'll have great difficulty justifying any discipline.
GRAED© #3	Make sure people know the policies (systems) and your expectations regarding their performance. Initial training is essential. Before putting a person into any job description, you must train her on how to do the activities she'll encounter and emphasize your performance expectations. You can't assume any new employee will know how to do an activity.

GRAED© #4	Make sure supervisors are behaving like supervisors. The primary mission of a supervisor is enforcement of organizational policy (systems). Someone has to assure that rules are being followed.
GRAED© #5	When rules (systems) aren't followed, someone needs to act. Discipline needs to be initiated whenever there's an observed or discovered deviation from organizational policy (systems). Any supervisor or manager who's aware of inappropriate, unsafe, illegal, or otherwise wrongful actions by an employee must act on this knowledge.
GRAED© #6	Take the time to think it through, and keep it private. Praise in public and discipline in private. That's a good rule to follow. It doesn't mean you allow unsafe or inappropriate behaviors to continue simply because other employees are present. You take the necessary action to stop the inappropriate conduct. Afterward, you take appropriate action against the employee in private after you've gathered your facts, talked to managers, human resources personnel, or your competent counsel.
GRAED© #7	Be specific when talking to the involved employee. You must present the policy, rule, or procedure violated and clearly state the violation specifics.
GRAED© #8	Keep it professional, fair, and balanced. If organizational discipline is going to have the intended goal of preventing future problems, it must be professional, fair, and balanced.
GRAED© #9	Documentation is essential. If it isn't written down, many will believe it didn't happen. Your documentation today will establish the record, justify the imposed action, and serve as a defense if any allegations are made against you for taking the action.
GRAED© #10	Get on with life and your organization's mission. Once the process is over, both the supervisor and involved employee must return to their respective jobs. Internalization of the issue by either party will only cause future problems.

Source: Graham Research Consultants, LLC.

Graham's Rules for Eliminating Sexual Harassment (GRESH©)

GRESH© #1	Ensure you have a comprehensive policy that's organizational specific, properly derived, and regularly reviewed by your competent counsel to confirm compliance in this rapidly changing area of state and federal law. Zero tolerance must be the rule. Make sure your policy includes a section on how to report and how to investigate deviations from your established policy.
GRESH© #2	Don't hire losers. Build your organization's commitment to the elimination of harassment into the hiring process.
GRESH© #3	Make sure your current employees know your organization's harassment policy. Knowledge is more than having a document saying you've reviewed the policy or been to a training session. Knowledge means you're regularly (at least quarterly) covering the policy and making sure each employee knows what it says. Harassment is a high-risk area and requires high-risk training.

GRESH© #4	Set the proper example. Personnel won't follow policy if supervisors, managers, and executives fail to comply with it.
GRESH© #5	Remember the principles of risk management. There are no new ways to harass people. Use the CATSINRO© analysis on a regular basis: Comments, Actions, and Things that are Sexual, Racial, Religious, or Sexual Orientation in nature which Reasonable People would find Offensive don't belong at work. Regularly inspect the workplace and look for problems.
GRESH© #6	Discourage supervisors and managers from dating subordinates. Dating subordinates is extremely imprudent. This situation is ripe for quid pro quo allegations and sets the wrong tone for the rest of the workplace.
GRESH© #7	Ensure you have a program for the prompt, fair, and impartial investigation of all complaints. Make sure you have a protocol for the standardization of investigation of claims. Most victims of harassment, whether it be sexual, racial, religious, sexual orientation, or any other protected class, don't want lawsuits or external investigations. They want the inappropriate behavior to stop. If they don't believe they'll receive an impartial investigation, they'll pursue external remedies.
GRESH© #8	Forget the welcome versus unwelcome analysis. What appears to be welcome behavior today can be considered unwelcome behavior tomorrow. How do you know the behavior was welcome? Zero tolerance must be the rule.
GRESH© #9	There's a definite link between the way your organization treats its employees, the way your employees treat each other, and the way your employees treat customers. You can't fix the outside until the inside is running smoothly.
GRESH© #10	Most things that go wrong in life are highly predictable. Treat people the way you'd like to be treated. If you don't have something good to say, don't say anything. Keep your hands to yourself. And you get more flies with sugar than with vinegar. The solution isn't complicated.

Source: Graham Research Consultants, LLC.

Graham's Rules for Improved Decision Making (GRIDM©)	
GRIDM© #1	Identify and clarify the activity. Ask what's going on in this activity, and what am I being asked to do?
GRIDM© #2	Is there discretionary time or not? There's no excuse for a poor decision when there's time to think the decision through. Failure to utilize discretionary time when available is overrepresented in subsequent tragedies.
GRIDM© #3	Am I able to address this activity? If the activity is within your current job description, then handle it now. If not, then immediately forward the activity to someone who can handle it, and follow-up to make sure it was properly handled.
GRIDM© #4	What's our current policy regarding the involved activity? What does our manual say about it? If there's a written rule, it has to be followed!

GRIDM© #5	What's our past practice regarding this activity? Use this discretionary time to ask someone who's done the activity before so that your behavior today is consistent with past practice.
GRIDM© #6	Is it the right thing to do under the circumstances? What are the ethical considerations of this activity? Every activity has an ethical concern, so every decision we make must include an ethical analysis.
GRIDM© #7	What are the potential consequences of my decision? You must understand and analyze potential consequences in advance of your decision. Consequences include intended, unintended, short-term, and long-term issues.
GRIDM© #8	Act! And if it involves preservation of life or property, act quickly. Make and implement your decision. If not a preservation of life or property issue, recognize it isn't too late to go back to step one in the process to assure you're still headed in the right direction. It's much easier to start over than it is to attempt to undo something that was done incorrectly.
GRIDM© #9	Document as necessary (that's the lawyer in me). Record-keeping and report writing are essential components in this process and must be done as activities develop. Don't think that you'll remember why you did, what you did, and when you did it three years ago.
GRIDM© #10	Learn from and share your experiences (that's the risk manager in me). If you learn something, share this new memory marker with your colleagues so all can benefit from your new knowledge.

Source: Graham Research Consultants, LLC.

Crisis Communication for Water-Related Entities

Understanding the Components of Media Relations (Designate, Appreciate, Accelerate, and Calculate)

Designate	Public Information Officer (PIO) • Designate a PIO, no matter how small your organization. • Develop a PIO team of trained individuals. • Those filling the PIO role should be smart, media savvy, and professional.
Appreciate	Politics/Relationships • It's always political. • How will you address political pressure? • Don't fight those who are against you. • Align with your supporters.
Accelerate	Work at the Speed of Social Media • You need to be fast and accurate. • New media creates new challenges and new options.
Calculate Media Relations = Policy + Training + Planning	Policy • Establish accountability. • Stress the importance of fast response. • Provide clear direction on who can speak to the media. • Create protocol for release of information on personnel and staff.

Training

- Establish formal training for your PIO (forty-hour course).
- Create informal training for all others who will/might talk with media (PIO trains others).
- Training topics should include news conferences, one-on-one interviews, on-camera etiquette, and general communications skills.

Planning

- Scheduled news conferences
  - Never speak extemporaneously.
  - Know what you can say and what you can't say.
  - Get your statement vetted.
  - Ensure all details are accurate.
- Impromptu interviews
  - Never say no comment.
  - Prepare and perfect a general message for these situations.
  - Know how to create a sound bite without saying much.

Source: Material credited to Graham Research Consultants, LLC, as well as Lexipol and their expert contributor, Frank Cowan.

Crisis Communication for Water-Related Entities

When the Headline Is You (A Review of Designate, Appreciate, Accelerate, and Calculate)

Few water/wastewater professionals cherish the media spotlight, and many can share encounters with reporters they later regret. That's why media relations and crisis communication remain an important topic for any essential service provider like water-related entities. The media has a powerful effect on your organization's community perception and goodwill. They also set the agenda, and their stories aren't always factually accurate or properly depicted. Since your role in the narrative has been precast, it's challenging to speak with the media when they're predisposed to a certain narrative. Let's delve deeper into the ways you can take back control. Talking to reporters during normal events is an unnatural dynamic that's replete with risk. These interactions become exacerbated when there's a controversial situation, organizational embarrassment, or catastrophic event involving your operations. In such situations, the media will ask aggressive questions, and that necessitates a purposeful crisis communication system comprising policy, training, and planning.

The following scenario accentuates the importance of a system that's deliberately designed, up to date, and fully implemented. Okay, it's 6:30 pm on a weeknight. You pull up to a scene of a catastrophic water main break impacting an entire neighborhood. There are millions of gallons of your water deluging homes and businesses. The main break has also damaged a local middle school and community hospital. The media is huddled outside the emergency incident center waiting to shout questions at you. Upon exiting the vehicle, you become enveloped in irritation and apprehension. This catastrophic event is active. You have the facts, but they're not entirely favorable to your organization. The failure of your infrastructure has impacted a large swathe of your customers, and people want answers and leadership. What do you do as you prepare for the ruck of reporters? Let's assume your organization has implemented a well-designed, up-to-date, and fully implemented crisis communication system comprising policy, training, and planning. As such, you're ready for the media blitz and aggressive questions. You quickly playback the six steps from your training.

First, you have a proper perspective of the situation. Your training taught you that most reporters are expected to provide audio, video, an online story, and a longer piece for print. You know the reporters will be rushing and looking for unambiguous themes such as good versus evil and portraying the people in their reporting as familiar characters like villain, hero, idiot, and witness. Context and perspective are common casualties in the reporting process, but you already know that through your policy, training, and planning. You also know emotion vanquishes fact in the court of public opinion. You won't be ignoring fact, but you'll absolutely be using emotion in an authentic and honorable manner. Second, you know immediate framing of the message is necessary to gain control of the narrative. Your message must show an appropriate level of concern for the effected community. Your fact gathering has confirmed the water main break that arose from employee negligence and infrastructure malfunction. Consequently, you commit yourself to portraying concern, transparency, and problem-solving so that future similar tragedies are prevented. Your message will also convey accountability, and you know the necessity of taking responsibility in a forthcoming way. The same applies to honesty and empathy as well as implementing a problem/solution formula where you'll use one sentence responses to identify the problem and solution. This formula allows you to penetrate the media edit. It similarly makes it more likely your comments on the solution won't be edited out of the story.

Third, you know not to answer in half-sentences because such quotes can be unclear and allow reporters to control the context by inserting their own words to introduce your quote. Fourth, you're already managing your breathing, as you know the significance in speaking clearly and deliberatively. Your training taught you that slowing down allows you to come across as thoughtful and measured. It also gives you more control over your words. Fifth, you remind yourself not to repeat the same message, as that tactic erodes trust and infers deceit. And sixth, you focus your mind to carefully think about the questions that will be asked of you. Concentration is essential because aggressive questions offer an opportunity to comment on deeper issues. You also know the importance of reframing question so your answers aren't defensive. With those memory markers in place, you exit the vehicle with a plan. Your nerves quickly dissipate as you revert back to your policy, training, and preparation.

Source: Material credited to Graham Research Consultants, LLC, as well as Lexipol and their expert contributors Connie McNamara, Shannon Pieper, and Jeff Ansell.

Crisis Communication for Water-Related Entities

The Importance of Perspective and Rapport in Media Relations (Know Their Name

A crisis communication system comprising policy, training, and planning will allow your organization to properly control media messaging. Such control is essential because the media has an inherently oppositional relationship with it subjects. This countervailing nature requires appropriate perspective and a media relationship built on mutual respect. Both professions serve the public and perform important services. Our Founding Fathers specified a free press in our Constitution, and the media plays an integral role in documenting history. The local media, who are waiting for you to exit your vehicle, don't intend to make your life miserable. They have a job to do and are understaffed. Equally important, you know their predisposition to simplify and condense context to make news stories interesting, compelling, and understandable. The following six steps provide additional perspective and underscore the importance of building meaningful relationships with local media.

Step #1: They aren't the enemy.

Most media are hard-working people who take their jobs seriously. The vast majority entered this field because they believe it's important, they care about their community, and they're passionate about telling stories.

Step #2: Help them get what they need, and they'll work with you.

Videographers and photographers need to capture images of the scene to effectively tell their story. If you allow them to get what they need, they'll be more likely to adhere and respect your guidelines. They'll work around you if they're blocked or prevented from taking good videos or photos. Grant them access, even if it's limited. If you can't provide access immediately, tell them you're working on it and when they can expect it. Treat them with respect, and they'll likely do the same for you.

Step #3: Never try to hide information or shut the media down.

Doing so will only make reporters more determined to get what they need and may infer you're doing something wrong. It doesn't help you to make the relationship adversarial.

Step #4: They report to a boss, and they're on deadline.

What may seem pushy to you is simply somebody trying to finish their assignment. Even if you can only give them a little information, that will help. Journalists have a limited time to file their stories. Small details are key to helping them tell an authentic story.

Step #5: Forgive the dumb questions.

They aren't the experts, so they may ask questions that you think are obvious. You work in one area, water/wastewater operations, whereas journalists may be at a crime scene one moment, your water main break another, a school board meeting the next day, and a complicated legal hearing after that. They need to know a little about a lot and have the ability to translate what you tell them into a language the average person will understand. Sometimes journalists ask simple questions because they know their viewers and readers need more explanation, and they know the message will be more compelling coming from your PIO.

Step #6: They can help if you keep the lines of communication open.

Do you need help finding who committed arson to your pump stations, or who's responsible for a rash of copper wiring theft from your electrical panels? Lean on the media to give community members a place to call or email if they have information that could help. You'll likely get better coverage if your PIO is interviewed on camera and has photos of the arson or copper wiring theft. Photos tell a more memorable story, which can increase your hotline tips. You'll have much more success with your local media relations if you commit to working with them. Once they know you're a straightforward communicator and responsive to what they need (as much as your position affords), they'll respect you and be easier to work with on events impacting the community.

Source: Material credited to Graham Research Consultants, LLC, as well as Lexipol and their expert contributors Connie McNamara and Shannon Pieper.

Graham's Rules for Improving Crisis Communication (GRICC©)	
How to Speak to the Media	
GRICC© #1	Develop a policy, a plan, and a team.
GRICC© #2	News and social media don't play by any sort of rules. Plan and act accordingly but within your policy and plan.
GRICC© #3	Build interview and communications skills through formal training.
GRICC© #4	Designate a Public Information Officer (PIO), and remember the following: Your PIO isn't the dumping ground for malcontents or sluggards.
GRICC© #5	Know when to use a prepared statement and when you can answer questions.
GRICC© #6	Develop a strong, cohesive message.

GRICC© #7	It's all political—deal with it.
GRICC© #8	Don't battle special interest groups.
	Focus on your supporters and allies.
GRICC© #9	Move at the speed of social media.
GRICC© #10	If you have time to think, use it.

Important Takeaway: Rules #9 and #10 aren't contradictory.

Source: Graham Research Consultants, LLC

The Ten "Fs" for Success

Maximizing Quantity and Quality of Your Life

Faith	Believe in a higher power.
Family	Take care of your family, always.
Friends	Invest in friends and acquaintances.
	Having two or three great friends who'll always be on your side is fantastic.
Fitness	Keep yourself in shape, just walking an hour a day is wonderful.
	Regular doctor visit, even if you're feeling great, is necessary.
	Always address your mental and physical concerns.
	Take care of yourself.
Food	Do everything in moderation, and follow this general rule: If your grandmother wouldn't recognize it, then don't eat it.
Fun	Laugh a lot.
	There are innumerable benefits to being happy and laughing a lot.
Funds	Prioritize financial planning early in your career and try to retire debt free.
	If you have something extra, give it to someone who really needs it.
	Follow the adage: it's better to give than to receive.
Freedom	Be grateful you reside in the greatness that's the United States of America.
	Protect the freedoms that so many have died for over the centuries.
	Never forget that freedom isn't free.
Future	Time flies by quickly.
	Strategic thinking is essential.
Fulfillment	Make every day count.
	Be humble: It's not all about you.
	You get the opportunity every day to make a difference in someone's life. Simultaneously, you're "building your dash."
	Take a look at the poem of that name by Linda Ellison.

Work hard, do the right thing, tell the truth, take every opportunity offered, make the call. I wish you continued success in all you're doing.

—Gordon Graham

Source: Graham Research Consultants, LLC.

Gordon Graham's Recommended Reading List

- Alan E. Diehl, *Silent Knights: Blowing the Whistle on Military Accidents and Their Cover-Ups.*
- Amanda Ripley, *The Unthinkable: Who Survives When Disaster Strikes—and Why.*
- Andy Brown, *Warnings Unheeded: Twin Tragedies at Fairchild Air Force Base.*
- Andrew Hopkins, *Disastrous Decisions: The Human and Organizational Causes of the Gulf of Mexico Blowout.*
- Atul Gawande, *The Checklist Manifesto.*
- Bruce Schneier, *Beyond Fear: Thinking Sensibly about Security in an Uncertain World.*
- Bryan Vila, *Evaluating the Effects of Fatigue on Police Patrol Officers.*
- Charles Perrow, *Normal Accidents: Living with High-Risk Technologies.*
- D. Michael Abrashoff, *It's Your Ship: Management Techniques from the Best Damn Ship in the Navy.*
- Dan O'Neill, *The Firecracker Boys: H-Bombs, Inupiat Eskimos, and the Roots of the Environmental Movement.*
- Daniel Kahneman, *Thinking, Fast and Slow.*
- Dennis Smith, *San Francisco Is Burning: The Untold Story of the 1906 Earthquake and Fires.*
- Dietrich Dörner, *The Logic of Failure: Recognizing and Avoiding Error in Complex Situations.*
- Douglas C. Waller, *Wild Bill Donovan: The Spymaster Who Created the OSS and Modern American Espionage.*
- Gary Klein, *Sources of Power: How People Make Decisions.*
- George Friedman, *The Next Decade: Where We've Been...And Where We're Going.*
- George Friedman, *The Next Hundred Years: A Forecast for the 21st Century.*
- James R. Chiles, *Inviting Disaster: Lessons from the Edge of Technology.*
- James Reason, *Human Error: Models and Management.*
- James W. Loewen, *Lies My Teacher Told Me: Everything Your American History Textbook Got Wrong.*
- Jeffrey K. Liker, *The Toyota Way: 14 Management Principles from the World's Greatest Manufacturer.*
- Jerome Groopman, *How Doctors Think.*

- Jody Hoffer Gittell, *The Southwest Airlines Way.*
- John Giduck, *Terror at Beslan: A Russian Tragedy with Lessons for America's Schools.*
- John Kasich, *Stand for Something: The Battle for America's Soul.*
- John McCain, *Hard Call: Great Decisions and the Extraordinary People Who Made Them.*
- Jonah Lehrer, *How We Decide.*
- Joseph T. Hallinan, *Why We Make Mistakes: How We Look Without Seeing, Forget Things in Seconds, and Are All Pretty Sure We Are Way Above Average.*
- Karl E. Weick, *Managing the Unexpected: Resilient Performance in an Age of Uncertainty.*
- Laura Hillenbrand, *Unbroken: A World War II Story of Survival, Resilience, and Redemption.*
- Laurence Gonzales, *Deep Survival: Who Lives, Who Dies, and Why.*
- Laurence Gonzales, *Everyday Survival: Why Smart People Do Stupid Things.*
- Malcolm Gladwell, *Outliers: The Story of Success.*
- Marc S. Gerstein, *Flirting with Disaster: Why Accidents Are Rarely Accidental.*
- Margaret Heffernan, *Willful Blindness: Why We Ignore the Obvious at Our Peril.*
- Mark Eberhart, *Why Things Break: Understanding the World by the Way It Comes Apart.*
- Matthew Parker, *Panama Fever: The Epic Story of One of the Greatest Human Achievements of All Time-the Building of the Panama Canal.*
- Michael D. Watkins and Max H. Bazerman, *Predictable Surprises: The Disasters You Should Have Seen Coming.*
- Michelle Wucker, *The Gray Rhino: How to Recognize and Act on the Obvious Dangers We Ignore.*
- Nicholas Taleb, *The Black Swan: The Impact of the Highly Improbable.*
- Nick Tasler, *The Impulse Factor: Why Some of Us Play It Safe and Others Risk It All.*
- Om Prakash Kharbanda, *Safety in the Chemical Industry.*
- Peter L. Bernstein, *Against the Gods: The Remarkable Story of Risk.*
- Peter M. Senge, *The Fifth Discipline: The Art & Practice of the Learning Organization.*
- Robert Cialdini, *Influence: The Psychology of Persuasion.*
- Sidney Dekker, *Just Culture: Balancing Safety and Accountability.*
- Spencer Johnson, *Who Moved My Cheese?*
- Stephen Wilson and Andrei Perumal, *Waging War on Complexity Costs.*
- Stephen Wilson and Andrei Perumal, *Growth in the Age of Complexity.*
- Steve Salerno, *Sham: How the Self-Help Movement Made America Helpless.*

- Steven Emerson, *Jihad Incorporated: A Guide to Militant Islam in the United States.*
- Thomas L. Friedman, *The World Is Flat: A Brief History of the Twenty-First Century.*
- Tim Harford, *Adapt: Why Success Always Starts with Failure.*
- Todd Conklin, *Pre Accident Investigations: Better Questions.*
- Tony Kern, *Blue Threat: Why to Err Is Inhuman: How to Wage and Win the Battle Within.*
- Tony Kern, *Darker Shades of Blue: The Rogue Pilot.*
- Tony Kern, *Flight Discipline.*
- Wikipedia, *Cognitive Lock.*

Index